ULTRAVIOLET LASER TECHNOLOGY AND APPLICATIONS

ULTRAVIOLET LASER TECHNOLOGY AND APPLICATIONS

David J. Elliott
Excimer Laser Systems
Wayland, Massachusetts

San Diego   New York   Boston   London   Sydney   Tokyo   Toronto

This book is printed on acid-free paper.

Copyright © 1995 by ACADEMIC PRESS, INC.

All Rights Reserved.
No part of this publication may be reproduced or transmitted in any form or by any means, electronic or mechanical, including photocopy, recording, or any information storage and retrieval system, without permission in writing from the publisher.

Academic Press, Inc.
A Division of Harcourt Brace & Company
525 B Street, Suite 1900, San Diego, California 92101-4495

United Kingdom Edition published by
Academic Press Limited
24-28 Oval Road, London NW1 7DX

Library of Congress Cataloging-in-Publication Data

Elliott, David, date.
   Ultraviolet laser technology and applications / by David Elliott
       p.   cm.
   Includes index.
   ISBN 0-12-237070-8 (alk. paper)
   1. Lasers. 2. Ultraviolet radiation. I. Title.
TA1677.E55   1995
621.36'6--dc20   95-20628

PRINTED IN THE UNITED STATES OF AMERICA
95 96 97 98 99 00 EB 9 8 7 6 5 4 3 2 1

Contents

Preface   xi
Acknowledgments   xv

1. Ultraviolet Light

1.1 Introduction	1
1.2 The Ultraviolet Spectrum	1
1.3 Historical Development of the Laser	3
1.4 How a Laser Works	4
1.5 UV Sources	10
1.6 UV Lasers	16
1.7 Inert Gas UV Lasers	21
1.8 Metal Vapor Lasers	22
1.9 Nitrogen Lasers	23
1.10 Helium-Neon Lasers	23
1.11 The Alexandrite Laser	24
References	26
Glossary of Terms	26
Reading Bibliography	30

2. Ablation

2.1 The Ablation Phenomenon	33
2.2 Background and History of Ablation	34
2.3 Early Discoveries and Uses of Ablation	35
2.4 Ablation Mechanism	36
2.5 Practical Characteristics of UV Reactions	46
References	54
Glossary of Terms	56
Reading Bibliography	58

3. The Excimer Laser

    3.1 Introduction   67
    3.2 History   67
    3.3 Theory of Operation   68
    3.4 The Gain Profile   70
    3.5 Excimer Laser Subsystems   71
    3.6 Laser Operation   75
    3.7 Laser Installation   82
    3.8 Laser Maintenance   82
    3.9 Gas Safety Guidelines   85
    3.10 UV Safety   86
    References   88
    Glossary of Terms   89
    Reading Bibliography   90

4. UV Materials Research

    4.1 Introduction   95
    4.2 Laser Mass Spectral Analysis   95
    4.3 157-nm Fluorine Laser Processing   98
    4.4 Production of X-Rays with 248 nm Energy   99
    4.5 Excimer Laser Induced Fluorescence   100
    4.6 Tunable UV Laser   101
    4.7 Laser Chemical Vapor Deposition   103
    4.8 Surface Analysis by Laser Ionization   104
    4.9 Photopolymer Research   104
    4.10 Laser Treatment of Organosilicons   107
    4.11 UV Laser Cleaning   107
    4.12 Cross-Sectioning Delicate Structures   110
    References   110
    Glossary of Terms   111
    Reading Bibliography   112

5. UV Optics and Coatings

    5.1 Introduction   123

CONTENTS vii

5.2 UV Optical Requirements	124
5.3 UV Optical Materials	125
5.4 Radiation Damage	126
5.5 Coating Manufacture	128
5.6 UV Coatings and Reflection	130
5.7 Extreme UV Coatings	131
5.8 Single Layer Antireflection Coatings	134
5.9 Multilayer Antireflection Coatings	136
5.10 Dielectric Reflector Coatings	138
5.11 Metal Reflector Coatings	140
5.12 UV Coating Types by Application	145
5.13 Coating Damage and Defects	151
5.14 Coating versus Polarization	156
5.15 Optical Fiber Beam Delivery	158
References	159
Glossary of Terms	161
Reading Bibliography	164

6. UV Laser Cleaning

6.1 Introduction	173
6.2 Technology Factors in Surface Contamination	174
6.3 Size of Contaminants	175
6.4 Contamination	177
6.5 Integrated Circuit Cleaning	177
6.6 Thin Film Heads	181
6.7 Flat Panel Display Cleaning	184
6.8 Compact Discs	187
6.9 Printed Circuit Boards	191
6.10 Cleaning Semiconductor Surfaces	196
6.11 Laser Ablative Cleaning	199
6.12 Debris Removal	201
References	205
Glossary of Terms	206
Reading Bibliography	207

7. Annealing and Planarizing

7.1 Introduction	209
7.2 Annealing Parameters	210
7.3 Laser versus Electron Beam	213
7.4 Pulsed versus Continuous Lasers	214
7.5 Annealing Process Control	215
7.6 Laser Planarization	220
7.7 IC Topography Problems	221
7.8 Resist Imaging on Topography	223
7.9 Planarization with Polymer Coatings	225
7.10 Nonlaser Thermal Planarization	226
7.11 UV Laser Planarization	226
References	244
Glossary of Terms	245
Reading Bibliography	246

8. Deep-UV Microlithography

8.1 Introduction	251
8.2 Evolution of Deep-UV Lithography	251
8.3 Deep-UV Technology: Resolution	254
8.4 Resolution: The ``k'' Parameter	254
8.5 Exposing Wavelength	256
8.6 Deep-UV Laser Wafer Stepper	257
8.7 Optical Materials for Deep-UV Imaging	259
8.8 Deep-UV Light Sources: Mercury Lamps	260
8.9 Line-Narrowed Excimer Laser Subsystems	261
8.10 Coherence	266
8.11 Measurement of Wavelength and Bandwidth	267
8.12 Laser Performance	269
8.13 Laser Maintenance	270
8.14 Utilities	272
References	272
Glossary of Terms	273
Reading Bibliography	277

CONTENTS

9. Micromachining

9.1 Introduction	287
9.2 Micromachining Processes	288
9.3 Direct Ablation	290
9.4 Micromachining Applications	295
9.5 Three-Dimensional UV Microforming	300
9.6 Microelectromechanical Systems	306
9.7 Failure Analysis Applications	308
9.8 Wire Stripping and Fiber Taps	308
References	311
Glossary of Terms	312
Reading Bibliography	313

10. UV Lasers in Medicine

10.1 Introduction	317
10.2 Applications	317
10.3 UV Laser Advantages in Medicine	318
10.4 UV Laser Angioplasty	320
10.5 Corneal Sculpting	325
10.6 Microbiology Research	327
10.7 Laser Microsurgery	329
10.8 Catheter Machining Optics	332
References	333
Glossary of Terms	334
Reading Bibliography	335

Subject Index 339

Preface

Ultraviolet laser technology is evolving from a research phase to commercial use in a wide variety of fields. Practical applications for UV laser energy have emerged in medicine, semiconductor processing, optical communications, micromachining, and other fields. Commercial deep-UV laser delivery systems are being manufactured to perform a variety of commercial products.

Increased activity and growth in UV technology are results of laser research technology being reduced to practice. The unique interaction mechanisms of UV laser energy with a wide variety of materials have been studied worldwide in many R&D labs, and published literature on ultraviolet laser phenomenon is rapidly expanding. UV laser technology provides unique functional features that include high resolution, athermal processing, and precise microstructuring and micropatterning capabilities. These practical capabilities have been recognized are now in demand in several fields.

Technical progress in UV optics, coatings, and UV laser source technologies has made it possible for the technology to expand to provide useful products. A summary of the fields where UV laser technology is being applied, specific applications, and key properties are listed below.

Field	Application	Key Property
Medicine	Corneal Milling	Athermal Reactions Surgery
	Microsurgery	Sharp, Precise Cuts
Semiconductors	Photolithography	Submicron Resolution
	Packaging	Direct Structuring of Metals and Plastics
	Cleaning	Ablation of Particle and Film Contamination
Optical Communication	Waveguide Fabrication	Precision Polymer Ablation and Micro-cutting
	Fiber Networks	Microstructuring of Glass and Plastic
Materials Research	Mass Spec Magnetics/Circuits	Ablation Reactions
Micromachining	Miniature Flow Motors/Devices	Direct Structuring by Ablation

The overall aim of this book is to provide a practical, "hands-on" reference source on the basic technical areas and applications of UV laser technology. This book is intended for researchers, technologists, process engineers, students, business managers, and newcomers to the field who require a basic introduction to the field. Each chapter deals with a different aspect of UV laser technology, beginning with UV light itself, moving through the optics, sources, and systems, followed by detailed descriptions of applications in various fields.

Chapter 1, "Ultraviolet Light," deals with the nature of UV energy and UV sources. Chapter 2, "Ablation", discusses the phenomenon of ablation, physics of the many reactions involved in ablation, and the by-products. Chapter 3, "The Excimer Laser," describes the primary light source for intense deep-UV radiation. Chapter 4, "Materials Research," reviews primary areas of UV laser research, such as spectroscopy, UV polymer research, and

UV biology. Chapter 5, "UV Optics and Coatings," covers the main optical materials and coatings for UV laser beam delivery. Chapter 6, "UV Laser Cleaning," deals with surface treatment using UV energy and chemical interactions. Chapter 7, "Annealing and Planarization," discusses the uses of UV laser energy for surface melting and planarizing in semiconductor technology; microlithography, another microelectronic application, is treated separately in Chapter 8. Chapter 8 on "Deep-UV Microlithography" describes the use of UV lasers and optics in a wafer stepper system used for imaging advance IC devices; Chapter 9, "Micromachining," explains the use of UV lasers for machining small structures in various materials; and Chapter 10 reviews medical applications of UV laser technology.

A glossary of terms and a reading bibliography are included at the end of each chapter.

Acknowledgments

I wish to thank a number of people who have contributed in different ways to the writing of this book, namely: Bernhard Piwczyk, whose vision and enthusiasm drew me into the field; Dr. Herbert Pummer, Jack Andrellos, Dr. Gerard Zaal, and Dr. Dirk Basting for introducing me to excimer lasers at Lambda Physik; Dr. Uday K. Sengupta, Bob Akins, Dan Wilson, and others at Cymer Laser Technologies for guidance and training on lithographic lasers; Glen P. Callahan and Dr. Bruce Flint of Acton Research for information on UV coatings; Dr. R. Srinivasan for information on ablation; Dr. Daniel J. Ehrlich of MIT, Lincoln Labs, for references to his pioneering work in the field; Dr. David C. Shaver of MIT, Lincoln Labs, for technical information and assistance in developing a deep-UV lithography tool; Dr. Bruce W. Smith of RIT for providing our first excimer laser; Dr. Richard F. Hollman, while Chief Scientist at Excimer Laser Systems and later at UV Tech Systems Inc., for many hours of cogitation and effort on deep-UV optical systems; Frank M. Yans and Dan K. Singer for working closely with me in the development of new deep-UV surface treatment technology at UVTech; Mr. George D. Whitten for continued friendship and inspiration; John Frakenthaler for help in computer searching, Amanda Peacock and Deanna Sklenak for their dilligent effort in preparing the manuscript; and the staff at Academic Press for their editing, technical assistance, and patience.

I also wish to thank the many authors, technicians, process engineers, managers, and other professionals in the field of UV laser technology whose data, papers, illustrations, references, and helpful discussions made the writing of this book possible.

Boston, March 1995
David J. Elliott

Chapter 1

Ultraviolet Light

1.1 Introduction

Ultraviolet light has become more important in recent years as the various technologies necessary to provide practical UV laser imaging and beam delivery systems have progressed, especially the light sources. Historically, UV light has been available only from relatively low power lamps, thereby restricting the usefulness of the technology. The discovery and development of the excimer laser in the 1980s made possible, for the first time, the availability of intense ultraviolet light. Researchers explored and uncovered the unique properties of this new light source. As various phenomena involving UV energy and material interactions were discovered and optimized, practical applications emerged. In this chapter, we discuss the UV spectrum, early UV lamp sources, laser physics and operating principles, and development leading to the discovery of the UV laser.

1.2 The Ultraviolet Spectrum

The "ultraviolet" is a small portion of the electromagnetic spectrum, yet within its boundaries there are three distinct regions (near, mid, and far UV), each with particular significance in terms of applications and properties. Ultraviolet light is roughly defined as that portion (400- to 100-nm wavelengths) of the electromagnetic spectrum between longer wavelength visible light (400-700 nm) and shorter wavelength x-ray (10-100 nm) energy. Figure 1.1 shows the ultraviolet, visible, and infrared spectrums, indicating the various emission lines (1).

CHAPTER 1. ULTRAVIOLET LIGHT

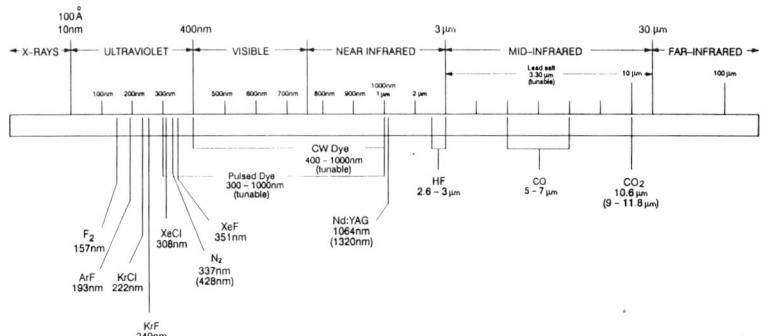

Figure 1.1: Laser Spectrum

1.2.1 Far or Vacuum UV

The shortest wavelength portion of the ultraviolet region of the electromagnetic spectrum is the "far UV" or "vacuum UV" (VUV), which is roughly the region from 100 to 200 nm. At these very shortest of the UV wavelengths, air becomes opaque, requiring that experiments be performed in a vacuum (or inert gas) so that the air does not absorb all the UV light. In commercial UV optical delivery systems, far or vacuum UV wavelengths must be contained in inert gas (argon, nitrogen) -purged beam containment tubes. If this is not provided, considerable energy losses will occur from air molecules absorbing the UV photons.

1.2.2 Deep-UV

The "deep-UV" portion of the ultraviolet spectrum is the region from approximately 180 to 280 nm, so-called because it is the deepest area of the ultraviolet where practical UV imaging is routinely done. It is also the deepest part of the UV spectrum where work can be done at atmosphere without side efforts.

1.2.3 Mid-UV

The "mid-UV" is the region of the ultraviolet spectrum from 280 to 315 nm, so-called because it is midway between deep-UV and near-UV.

1.2.4 Near-UV

The "near-UV" is the region from 315 to 400 nm, so-called because it is nearest to the visible portion of the electromagnetic spectrum.

1.3 Historical Development of the Laser

Before UV lasers, there were long wavelength, red lasers. In this section, we will discuss the origins of the UV laser back to the initial concepts of amplification of wave energy. The origins of the laser (<u>L</u>ight <u>A</u>mplification by <u>S</u>timulated <u>E</u>mission of <u>R</u>adiation) are traced back to the early 1900s when Albert Einstein, Niels Bohr, Max Plank, and Ernest Rutherford were developing theories about the nature and behavior of matter.

Rutherford formed the idea of the atom composed of a nucleus with orbiting electrons. The electron orbits occurred at various energy levels. Plank developed the idea of electromagnetic waves as the form taken by radiant energy, and further specified that each frequency had a fixed quantum or amount of energy.

Niels Bohr described the phenomenon of fluorescence or spontaneous emission. This occurs when an atomic electron drops from a high energy level to an unoccupied lower energy level. When this happens, a quantum of light is emitted spontaneously.

Einstein theorized that other forms of emission were possible and formulated the idea of stimulated emission. Einstein predicted that when energy was applied to atoms, the response would be emission of energy.

These scientific theories remained to be proven in the lab, and during World War II, another step was taken to identify what became known as "lasing action" or lasers. High frequency radiowave and microwave oscillators were developed that could generate electromagnetic waves of very high frequencies and short wavelengths. The RADAR (<u>R</u>adio <u>D</u>etection <u>A</u>nd <u>R</u>anging) was a useful result of this work. The <u>M</u>icrowave <u>A</u>mplification by the <u>S</u>timulated <u>E</u>mission of <u>R</u>adiation (MASER) was also developed before the laser and proved the theory of population inversion.

An analogy of how light behaves in a laser can be taken from how sound behaves in an amplifier. If a loudspeaker is placed too close to a microphone, the sound from the speaker is reamplified through the microphone and produces a howl. The howl is caused by sound waves, placed at closely spaced intervals, reinforcing each other by being "in phase." At a certain level, an oscillation begins which produces the howl (3). In the following section, we will explain how this same principle works with light to produce a laser beam.

1.4 How a Laser Works

Lasers are possible because of the way light interacts with atoms and molecules. To describe how a laser works, we will first review some basic aspects of light/matter interaction. The electrons in an atom or molecule exist in very specific energy levels, called "states." Each atom possesses electrons that are characteristic to the specific element, or combination of elements (molecules) represented.

1.4.1 Energy State Transitions

When an electron moves from one state to a lower energy level, or state, the atom gives up the excess energy as a "photon" of electromagnetic radiation (either light or x-rays). The amount of energy carried away equals the difference in the energy between the original higher state and the new lower state. In molecules, where these movements (transitions) between states involve motions of entire atoms rather than single electrons, the same phenomenon generally produces lower-energy photons (infrared light).

1.4.2 High and Low Energy Photons

The energy carried by a photon is determined by how rapidly the light waves in it oscillate, and this oscillation is measured either as the frequency (number of oscillations per second) or wavelength (distance that the waves move during one oscillation). Wavelength and frequency are related to each other by the speed at which the photons (waves) move—the speed of light. The energy of a photon (or, equivalently, the frequency or wavelength) determines its color. Blue light photons have more energy–a higher frequency and a shorter wavelength–than red light photons.

1.4.3 Spontaneous Emission

An electron in a particular state can also absorb a photon that has an energy equal to the difference between that state and a higher one. As a result, it can jump to the higher one (called an excited state), where it stays for a period of time before giving up energy by radiating a photon of the same or

1.4. HOW A LASER WORKS

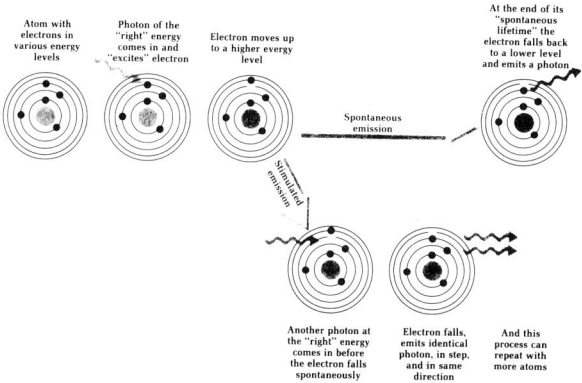

Figure 1.2: Schematic of the Principle of Stimulated Emission

different energy, and returning to some lower state. This time is called its "lifetime for spontaneous emission." Spontaneous lifetimes vary enormously, but are typically between a thousandth and a billionth of a second for levels of interest in common lasers. A spontaneously emitted photon can come out in any direction with equal probability.

1.4.4 Stimulated Emission

When Einstein recognized the principle of spontaneous emission, he showed that since a lower state electron in an atom can absorb a photon of the proper energy and go to a higher state, then one already in that higher state must have an equal probability of emitting a photon and dropping to the lower state when a matching photon comes past it. The process of creating emission by outside stimulation (as opposed to it occurring spontaneously) is called "stimulated emission." Further, this emitted photon must travel in exactly the same direction and with exactly the same energy as the photon that stimulated its emission (Figure 1.2).

1.4.5 Amplification

Therefore, if a large number of matching-energy photons comes past an electron in an excited state, there is a high probability that this process occurs so that its lifetime for stimulated emission can be much less than its spontaneous lifetime. Stimulated emission was reduced to practice when Charles Townes and others saw how to amplify (add photons to) a weak beam of light, with the addition consisting of photons that match the transition between energy states in some atom or molecule. This amplification is the critical process that makes a laser possible.

In designing a laser, the first major challenge is to devise some scheme in which a rather large number of excited atoms or molecules collect in an upper energy state. This alone, however, does not guarantee lasing action. Under most circumstances, electrons excited by an external energy source come to a normal condition called "thermal equilibrium." In the absence of any outside influence, the electrons in an atom or molecule will fill up the lowest of the many allowed energy states. If some outside source does supply energy (for example, photons with matching transitions or high-energy electrons that force others into a higher energy level), then some electrons will jump into higher energy states and drop from level to level until they reach the lowest unoccupied energy state, emitting a photon for every transition, and ending up in the normal condition. This means that in a population of atoms in the normal condition, there are always fewer electrons at high energy than at low energy.

1.4.6 Population Inversion

Because there are more atoms with electrons in the lower energy state of a transition, the absorption of photons by the electrons in the lower state is always greater than the stimulated emission of photons due to electrons in the upper state and there is always a net absorption (loss) of photons (1).

Therefore, the second major challenge in designing a laser is finding a way of decreasing the population of the lower energy state of a transition while maintaining the population of electrons in the higher state. This is called a "population inversion" since it is the inverse of what one would usually expect for a medium. A population inversion will result in little or no absorption of matching photons but with a large amount of stimulated emission from a higher state.

Researchers analyzing different materials have found thousands of transitions where it is possible to achieve some degree of population inversion, but there are relatively few that give inversions both easily and efficiently. Lasers in practical applications are usually selected from this latter category.

1.4. HOW A LASER WORKS

1.4.7 The Gain Medium

The first property that most people associate with a laser is its highly directional beam that does not spread as it propagates, but it is important to note that this is not necessarily so. The material in which the population inversion was created, called the "laser gain medium," will amplify photons moving in any direction. If we allow photons to go in any direction, the directionality (beam quality) of the laser will be no better than any other light source. The high directionality of the laser does not come from the laser medium itself, but from another critical component, the laser resonator or "cavity."

A simple laser resonator has two flat mirrors facing each other at each end of the laser gain medium and an aperture that restricts the diameter in which photons can propagate. Suppose an atom in the laser medium spontaneously emits a photon travelling at a wide angle to the resonator axis. The principle of the gain medium in a UV laser is illustrated in Figure 1.3. This photon bounces between the mirrors and may make some trips through the laser medium, but eventually it will "walk out" of the resonator, then both the original photon and the few additional photons whose emission it may have stimulated from the atoms in the laser medium will be lost.

Suppose, however, that an atom spontaneously emits a photon that travels exactly perpendicular to the two mirrors. This photon and the additional ones produced by stimulated emission never "walk out" of the resonator, they just bounce straight back and forth. A very intense beam of photons moving exactly perpendicular to the two mirrors builds up by the stimulated emission process and each pass through the gain medium. If we let some of this light out of the resonator, usually by making one of the mirrors slightly transparent, a laser beam emerges that has very high directionality or "beam quality" and a precise color determined by the atoms or molecules used for the gain medium.

Therefore, to obtain a highly directional output beam from the laser, the laser designer chooses a laser resonator with a low gain per pass (so that many passes are required to build up to a reasonable intensity) and a small-diameter aperture. However, under these conditions, it is difficult to get very high output power from the laser. Therefore, very large lasers often use a small, low-power laser "oscillator" to establish a beam with good quality. They then amplify the weak beam by passing it through a large, high-gain amplifier containing the same laser medium. Large laser systems for inertial fusion and laser isotope separation use this so-called "master-

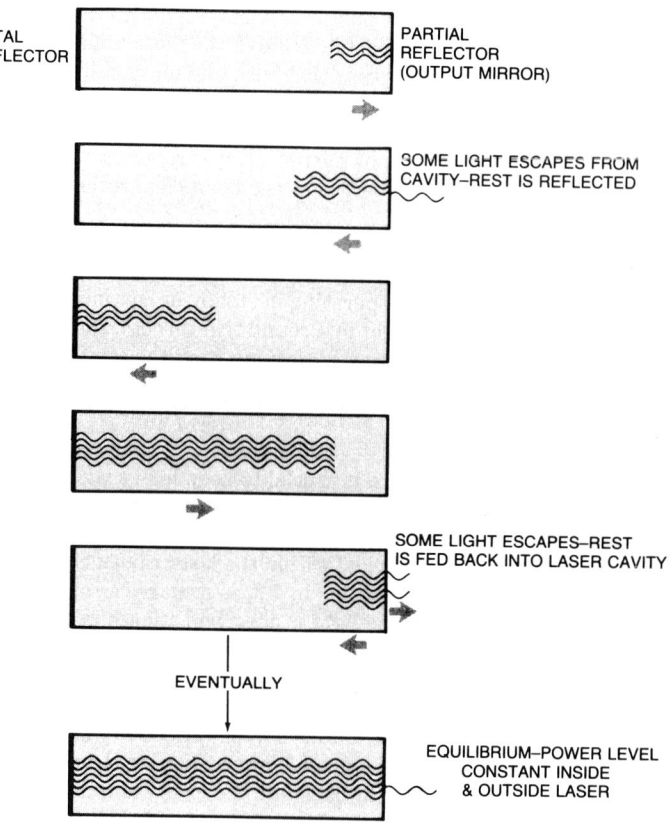

Figure 1.3: Principle of a Laser Resonator

1.4. HOW A LASER WORKS

oscillator/power amplifier" architecture.

1.4.8 Summary

In summary, for laser operation, three basic conditions must be met. First, there must be a suitable energy source to excite atoms, molecules, or ions of the active medium to a higher energy level. When they drop back to lower energy levels, photons of energy are emitted.

Second, there must be an active medium (of atoms, molecules, or ions) that will emit photons when stimulated. Third, there must be some form of optical feedback, or gain, generally provided by mirrors in the laser's optical cavity. This produces resonance.

The energy source can be provided from an electric current, a lamp, or by a chemical reaction. The material that emits laser radiation can be a gas medium, a liquid, or a solid, such as a semiconductor. These materials range from semitransparent crystals and glass to gases, vapors, and liquid held in a transparent vessel or circulated or sprayed into the laser cavity. The reflecting mirrors at both ends of the cavity which hold the active medium allow the optical radiation to resonate back and forth.

The energy source pumps or excites the active medium to a higher than normal energy state. High energy states are short-lived. Laser action begins when there is a greater amount of active medium at a higher energy state than at a lower state. This is called "population inversion." When some of the high-energy electrons fall back to normal levels, photons of light are released that stimulate the release of more photons in an increasing cascade, which is being reflected back and forth between the mirrors. When there is sufficient feedback in the optical resonating system, oscillation occurs and the medium begins to lase.

A laser beam can generally be effective only when it leaves the cavity. The laser beam cannot escape if both mirrors are 100% reflective. Further, no lasing can occur if either mirror is not reflective enough. In order to allow some of this energy to leave the cavity (to form a useful beam), a mirror that is not totally reflective must be installed at one end. This mirror is known as the "output coupler" or "transmitter." The optimum ratio of transmitted versus intracavity energy is not constant for all lasers, though in every case most of the optical energy must remain within the optical cavity to sustain laser oscillation.

These basic principles apply to lasers of all wavelengths, from far infrared through deep ultraviolet and into x-rays. We will now focus on the ultraviolet light in general, beginning with lamps and leading into lasers that emit various amounts of UV light.

1.5 UV Sources

Ultraviolet light is available from a variety of sources, including naturally occurring UV light, UV lamps, and UV lasers. UV light from the sun is abundant, but cannot be easily or economically harnessed for the applications that have practical value. To a lesser degree, UV lamps have the same problem: they supply rich amounts of the ultraviolet wavelengths, but by the time the energy is collected and transmitted through an optical system, very little energy is left to meet the demands of most commercial applications.

UV energy from lasers, however, solves this problem, providing a variety of wavelengths that are useful at intensities sufficient to perform economically viable applications. Since UV lamps have preceded lasers in many of these applications, we begin with a discussion of their attributes and contributions to UV applications and technology.

The main sources of UV light are as follows:

- Mercury Lamps (doped/undoped)

- Deuterium Lamps

- Germcidyl Lamps

- Argon Ion Lasers

- Excimer Lasers

- Alexandrite Lasers

- Tripled and Quadrupled Nd:YAG Lasers

- Fluorine Lasers

- Helium Cadmium Lasers

- Metal Vapor Lasers

- Nitrogen Lasers

1.5. UV SOURCES 11

1.5.1 Deep-UV Lamps

Deep ultraviolet radiation can be supplied from several nonlaser or lamp sources. Deep-UV lamps are typically used in applications that do not require the high average power levels available from excimer and other UV lasers. Deep-UV lamps are also relatively inexpensive and simple to integrate in an exposure system. In this section we will review the more common types of deep-UV lamps and their characteristics and describe typical optical delivery systems for practical applications.

Several sources are used for deep-UV lamps, each selected mainly for matching the emission spectrum as closely as possible to the response characteristics of the substrate or coating being exposed. Sources are also selected according to their conversion efficiency and output power.

Deuterium produces a continuum in the 200- to 315-nm spectral region, and deuterium lamps with up to 300-W output are useful tools for deep-UV research applications. Early use of deuterium lamps was to expose thin photopolymer coatings for photoresist research.

Mercury is the most common source for deep-UV lamps. Mercury sources deliver relatively high output powers (200-2000 W) in useful spectral regions. Further, mercury lamps can be "doped" or given additives to shift both the spectral output (wavelength) and the power at a given wavelength. High pressure, short arc mercury lamps are widely used for many industrial and research applications. Increasing the pressure inside the lamp envelope provides a means for higher output power. Figure 1.4 shows the emission spectrum of a xenon-mercury source (4).

High pressure mercury capillary lamps are used for very short UV applications. Mercury lamps are also pulsed by being driven from their idling power to very high intensities to produce short, high intensity pulses.

1.5.2 Cadmium-Doped Lamps

Low pressure, long arc cadmium-doped lamps offer high conversion efficiency. The cadmium dopant is responsible for high spectral output in the 215- to 235-nm region. Figure 1.5 shows the spectral output of a cadmium-doped lamp (4).

Dopants such as cadmium must be kept in vapor form to be effective, and heating of the source envelope is often required for maximum energy output. Microwave radiation is one technique for exciting and heating (pumping) lamp dopants. Most high-energy output lamps run very hot, and both air

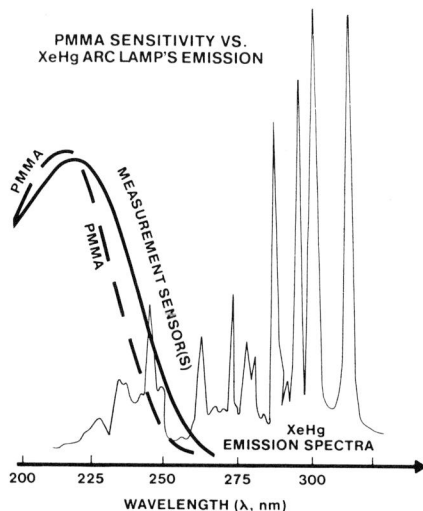

Figure 1.4: Xenon-Mercury UV Lamp Emission Spectrum

and water cooling are used to isolate thermal gradients from the collection and light delivery optics.

1.5.3 Pulsed Xenon

Pulsed xenon is a source capable of producing short wavelength deep-UV radiation in the 200- to 260-nm spectral range. When pulsed at high energy (1200 J) and a short pulse width (< 50 μsec), very high current density is achieved with relatively good (6-12%) conversion efficiencies. Various lamp shapes (elbow, circle) can be made to optimize the collection of energy. Pulsed xenon lamps produce high levels of RFI which must be shielded from the electronics of exposure systems.

1.5.4 Lamp Optical Delivery Systems

The purpose of the optical delivery system is to transmit a maximum amount of the UV energy to the substrate, to collimate the energy, and to eliminate heat transfer. A simple example of a UV lamp delivery system is shown in Figure 1.6.

This delivery system makes good use of mirrors and has only one refrac-

1.5. UV SOURCES

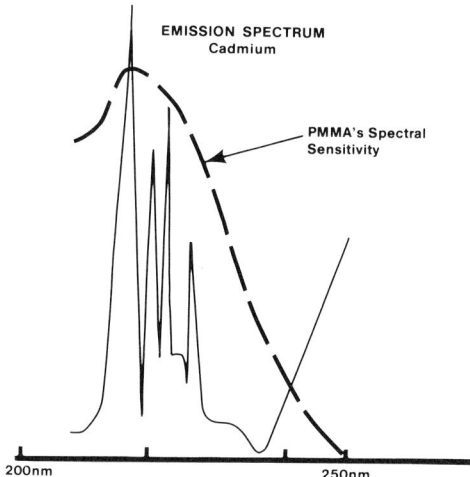

Figure 1.5: Spectral Output of a Cadmium-Doped Lamp

tive element, a collimating lens. Flat mirrors are used to surround the lamp, and a dielectric mirror (first one) reflects the UV through the optical axis to the substrate and transmits the infrared (heat) energy out of the imaging path.

Another example of a deep-UV lamp optical delivery system is shown in Figure 1.9. This systems uses an elliptical collection with the source at the first focus. A second focus point occurs at the multielement homogenizer or optical integrator point. The energy or rays become uniform after this point and are further highly collimated by the last refractive element, the collimating lens.

The main improvement in this system over the previous one shown is higher collection efficiency, which will produce higher energy intensity at the substrate plane. Also, the added refractive elements produce higher collimation and better beam control. The lamp is placed at the first focus of the elliptical reflector which collects 75% of the emitted radiation. The second focus of the ellipse is at the point of the optical "scrambler." The design of the scrambler or homogenizer is such that the user can predict and control optical divergence. A well-collimated, uniform beam of UV energy is thereby delivered to the wafer or substrate plane.

All of the mirrors used in these optical delivery systems are coated with

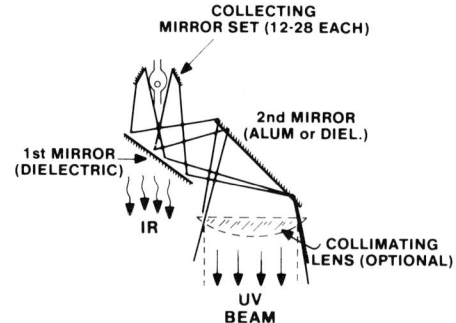

Figure 1.6: UV Lamp Optical Delivery System

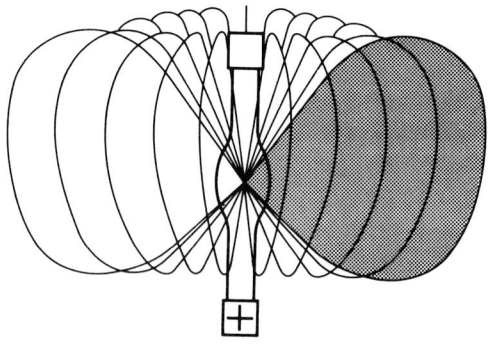

Figure 1.7: Three-Dimensional View of the Output of a Mercury Arc Lamp

1.5. UV SOURCES

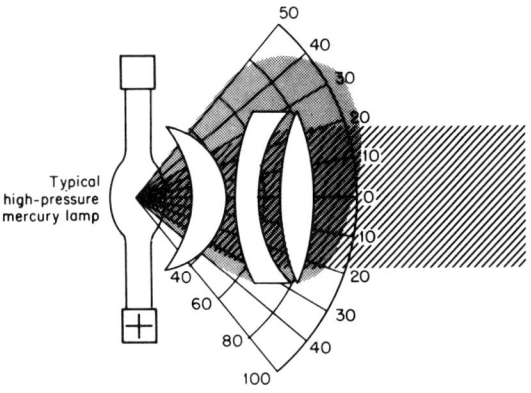

Figure 1.8: Lamp Illuminator Optics

special UV-reflecting materials which have a high index of refraction and $> 90 - 95\%$ UV transmission.

1.5.5 High Efficiency Lamp Delivery Optics

Some lamp-based exposure systems achieve very high intensity by a combination of electrodes, high efficiency collection optics, and special beam delivery optics. Figure 1.7 shows the three dimensional view of an output of a mercury arc lamp. FIgure 1.8 shows illuminator collection optics with relative intensity distribution which is then sent to the optics of the delivery system, typically a multiple element fiber bundle used to finally deliver the UV light to the substrate (see Figure 1.10).

The lamp used in this example is used for UV exposure of photoresist materials. The lamp is a 1000-W capillary mercury arc with a pressure of 230 atmospheres, containing two electrodes. The lamp envelope is cooled by jets of air to a temperature low enough to eliminate spectral output changes.

1.5.6 Lamps vs Lasers

Lamps as sources for deep-UV energy have a low cost and have a relatively long life with stable output and little or no maintenance. The main limitation of lamps (compared to lasers) is low output at deep-UV wavelengths. For example, deep UV micromachining requires photoablation of materials into

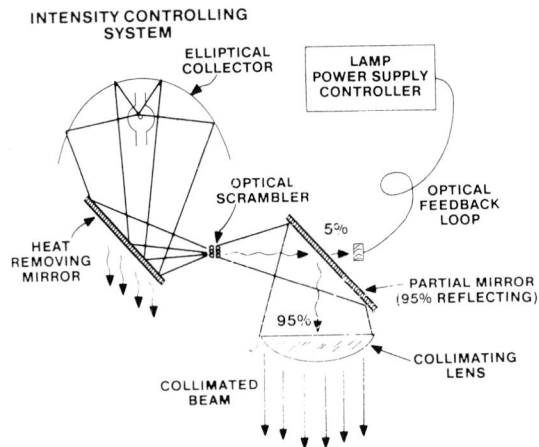

Figure 1.9: High Efficiency Deep-UV Optical Delivery System

largely molecular by-products. Lamps do not have sufficient power to initiate photoablation reactions.

Lamps have been used for many years in applications requiring low to medium UV intensity. New applications have emerged for high intensity UV sources that only excimer lasers can satisfy. Excimer lasers, discussed in detail in Chapter 3 (The Excimer Laser), provide the most intense source of UV light.

UV lamps continue to be used commercially for many applications. Examples include exposure sources for graphic art films, "black light" sources for identifying fluorescence in minerals, photoresist exposure sources (at 365 nm) for IC manufacturing, street lamps, sun tanning lamps, antibacterial sources for use in water purification and food processing, and copy machines.

1.6 UV Lasers

UV lasers are used as a source of ultraviolet light for at least two main reasons: The application demands an energy flux available from a UV laser (lamps not sufficiently powerful), or the properties of the UV light are ideally suited (optimal wavelength coupling) to the application, such as in microstructuring with a high resolution beam. In this section we will discuss UV laser sources, which include excimer lasers, argon, argon-krypton, kryp-

1.6. UV LASERS

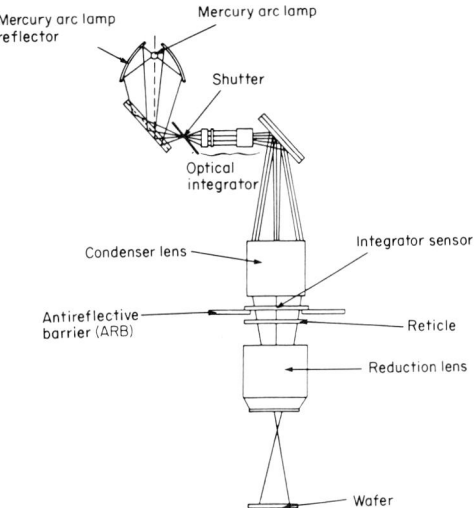

Figure 1.10: UV Lamp Optical Delivery System

ton lasers, helium-neon lasers, frequency-multiplied YAG lasers, nitrogen lasers, and Alexandrite lasers.

1.6.1 Excimer Lasers

<u>Theory of Operation</u>: Excimer lasers are a family of pulsed lasers operating in the ultraviolet region of the spectrum. The source of the emission is a fast electrical discharge in a high pressure mixture of a rare gas (krypton, argon, or xenon) and a halogen gas (fluorine or hydrogen chloride). The particular combination of a rare gas and halogen determines the output wavelength, and in most commercial excimer lasers, the wavelength may be changed by refilling the laser with the appropriate gas combination. Available wavelengths include:

- Argon Fluoride (ArF) 193 nm

- Krypton Fluoride (KrF) 248 nm

- Xenon Chloride (XeCl) 308 nm

- Xenon Fluoride (XeF) 351 nm

For users with unique requirements for pulsed radiation in the vacuum ultraviolet, a further output arising from the molecular fluorine (F_2) transition at 157 nm is available, although the output on this gas mix is weaker than any of those just listed.

Typical Performance: Typical outputs on the stronger transitions are pulse energies of a few hundred millijoules to 1, and repetition rates of 10 Hz to a few hundred hertz. In some cases, repetition rates of up to 1 kHz are available. Average powers are normally in the 10- to 100-W range. The pulse length of the excimer lasers is short, typically in the 10- to 30-nanosecond (nsec) range. This results in peak powers of several tens of megawatts.

Applications: The combination of short ultraviolet wavelengths with high peak and average powers makes excimer lasers uniquely suited for a wide range of applications.

Excimer lasers are used as pumps for high power tunable dye lasers, plasma studies, raman shifting, nonlinear studies, VUV generation, and photolysis.

In materials and semiconductor processing, applications include photolithography, chemical vapor deposition, annealing of semiconductors, fabrication of high-Tc superconducting films, high resolution machining of plastics and metals, and IC packaging by ablation of polymers.

Excimer lasers are also used in marking applications on wires, cables, ceramic, glass substrates, and on plastic components.

In medicine, applications exist for corneal surgery, laser angioplasty, endartarectomy, lens (intraocular) making, neurosurgery, and bone machining.

In environmental science, applications are fluorescence studies, remote sensing, and ozone monitoring. Many of these applications are covered in detail in various chapters of this book.

A detailed description of the excimer laser is given in Chapter 3, "The Excimer Laser."

1.6.2 UV Nd:YAG Lasers

The neodymium-YAG (Nd:YAG) laser is used for many industrial applications in aerospace, medical, automotive, and semiconductor industries. The long pulse duration and basic 1064-nm wavelength produces thermal overload in many metals and nonmetals. This fundamental interaction causes melting and vaporization reactions to occur, and is the basis for using Nd:YAG lasers for welding, cutting, drilling, trimming, marking, and many other practical operations used in manufacturing.

1.6. UV LASERS

The application of the Nd:YAG laser to ultraviolet applications is possible because of the high peak power of this laser, generating sufficient photon density inside a frequency multiplying crystal for efficient doubling (532 nm), tripling (355 nm), and quadrupling (266 nm) of the fundamental 1064-nm wavelength.

The Nd:YAG laser is highly reliable, very stable, and easy to use, and hundreds of these lasers are performing as "workhorses" in the volume production of many products. In this section we will describe the construction of the UV Nd:YAG laser and discuss harmonic generation for UV applications, diode pumping, 266- and 355-nm characterization, and safety.

Laser Construction: The Nd:YAG laser is constructed from a few basic subsystems. These include: a) Nd:YAG rod; b) a pump chamber; c) flash lamps; d) resonator optics; e) a "Q" switch; f) power supply; g) support electronics (pulse forming network, Q=switch driver); and h) water cooling. The schematic in Figure 1.11 shows the basic subsystems of the Nd:YAG laser.

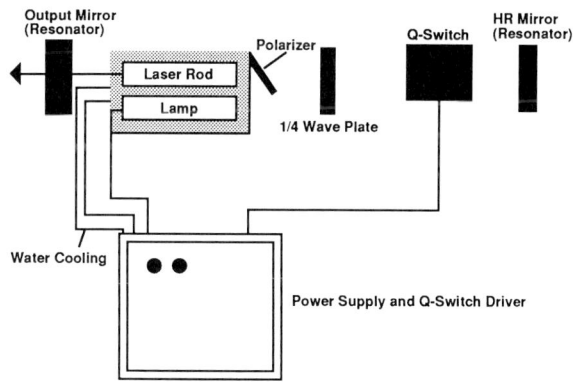

Figure 1.11: UV Nd:YAG Laser System Schematic

Operating Principle: The flash lamps in UV Nd:YAG lasers are powered by a power supply pulse forming network (PFN). Light from the flash lamps is used to pump the Nd:YAG rod. The pump chamber geometry is configured so as to condense and focus the flashlamp light efficiently into the rod. The Nd:YAG rod stores the flashlamp energy in the form of neodymium atoms excited to the lasing level. Excitation (submillisecond) is followed by spontaneous lasing.

Optimizing the storage of energy in the rod is accomplished through the use of a Q-switch, an electro-optic device made from a potassium-dideuterium phosphate (KD*P) crystal. The KD*P material rotates the polarization vector of the light beam when an appropriate electron field is applied.

The Q-switch prevents oscillation and allows excited atoms in the rod to build to a significant number. The opening of the Q-switch releases the energy stored in the rod, resulting in a short (5-6 nsec), intense pulse. The short pulse length and high energy provide relatively high peak power for lasers of this type. The natural 1064-nm wavelength is sent into a harmonic generator to produce both visible and/or UV beams (6).

Harmonic Generation: Optical crystals are used to produce doubled, tripled, and quadrupled frequencies of the fundamental wavelength. The second harmonic (doubled) is 532 nm, the third (tripled) harmonic is 355 nm, and the fourth (quadrupled) harmonic is 266 nm. Efficient frequency multiplication is achieved since the Nd:YAG laser produces relatively high photon densities.

The KD*P crystals are good harmonic generators because of their high nonlinearity, excellent UV transmission, and high damage resistance. The high figure of merit for nonlinearity (d constant) for frequency conversion is responsible for the change in the light passing through the crystal as a function of its orientation to the incident laser beam. The main nonlinearity property is birefringence, where the horizontally and vertically polarized light travels at different speeds in the KD*P crystal. This property is responsible for harmonic generation.

An example of the sequential mechanisms that occur in frequency doubling of the Nd:YAG is as follows: Two photons exit the laser at 1064 nm and enter the KD*P crystal close together, where they are combined into a single photon. They enter with the same circular polarization. The high efficiency of the crystal and the exiting doubled photon of doubled frequency achieve energy conservation.

Angular momentum conservation, also required for this process to be practically useful, is assured by horizontal polarization of the combined output photon. This polarization is achieved by using the crystal at the phase match angle (angle at which the crystal's birefringence equalizes the velocities of the input photons and the orthogonally polarized combined output photon) (6).

The second harmonic (532 nm) exits the crystal along with some portion of the 1064-nm natural wavelength. Generation of the third harmonic (355

1.7. INERT GAS UV LASERS

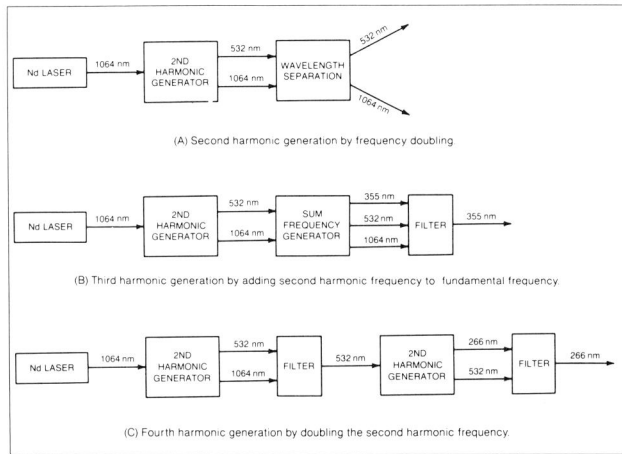

Figure 1.12: Third and Fourth Harmonic Generation with a Nd:YAG Laser

nm) or tripling occurs when a 1064-nm photon combines with one of the doubled photons. Since the doubling crystal has both of these photons, the twin beams are sent into another KDP crystal which combines them into a single, tripled photon of 355 nm length, the frequency sum of two mixed photons. The schematic in Figure 1.12 shows this example, along with an example of quadrupling to 266 nm.

The last wavelength of interest, and the one with the highest photon energy, is 266 nm, the fourth harmonic. Quadrupling the 1064-nm laser is achieved by using two doubling crystals in tandem. The first crystal produces 532-nm and some 1064-nm light; the 1064-nm beam is removed and the 532-nm beam is sent into another doubling crystal, producing 266-nm light (6). The frequency multiplied Nd:YAG lasers are finding use in applications where fairly intense UV energy is needed, somewhat below the power levels achieved with standard (20-200 W) excimer lasers.

1.7 Inert Gas UV Lasers

There are several types of inert gas lasers that can produce continuous beams of light at wavelengths ranging from the ultraviolet through infrared. They are based on a mechanism where an electrical circuit is discharged through the gas medium causing inert gas atoms to excite, then decay and simultaneously emit photons. The gases used are all elements of the rare gas (Group III) column of the periodic table. The most common of the inert gas lasers are argon, krypton, argon-krypton, and helium-neon.

1.7.1 Argon and Krypton Ion Lasers

Argon and krypton, both rare and inert gases, are used to produce laser energy at several wavelengths. Argon and krypton, like helium and neon, are also excited by discharging an electrical current through them, which provides the necessary energy for lasing action.

Argon and krypton lasers operate on the same principle as other inert gas lasers. Argon lasers produce continuous beams of light at wavelengths ranging from blue to green at power levels up to about 100 W. They are the most powerful of the visible light gas lasers. Krypton lasers produce light from blue to red, but cannot generate the power levels that argon can. When combined, argon and krypton are used to produce laser light at several visible wavelengths (all the wavelengths generated by either gas separately). Because of the wide range of colors they produce, argon and krypton lasers are used in light shows (5).

1.8 Metal Vapor Lasers

Metal vapor lasers use mixtures of metal vapor and rare gases to produce laser light. They operate at high temperatures (200-1200°C) to keep metals in a vapor state and can be excited by electrical discharges in mixtures of the metal vapor and rare gases. They produce light in the visible, ultraviolet, or infrared regions of the spectrum.

1.8.1 Copper Vapor Lasers

Copper vapor lasers produce green and yellow light from a mixture of copper vapor and helium or neon. They are excellent sources of short, high-intensity laser pulses at a very high pulse-repetition rate. Copper vapor lasers have a reported efficiency of 1% (electrical energy to light output or "wall-plug" efficiency), substantially better than the helium-neon or argon lasers. Copper vapor lasers produce up to 50 W of average power. Copper laser arrays have been scaled to very high average power levels for isotope separation in which they act as pumps for tunable dye lasers. There are also medical applications for the copper vapor laser.

1.8.2 Helium-Cadmium Laser

The helium-cadmium laser, a common source of ultraviolet light, operates by generating electrically excited helium atoms which transfer their excess energy to a vapor of cadmium. Cadmium is a soft, bluish metal which vaporizes at several hundred degrees centigrade, or a lower temperature than other metal vapor lasers. The helium-cadmium laser is therefore easier to engineer and manufacture, a fact which increases both its practical applications and overall commercialization in industry.

Helium-cadmium lasers produce a small (few hundredths of a watt) amount of ultraviolet light, and applications include typesetting and materials research.

1.8.3 Gold Vapor Lasers

Another type is the gold-vapor laser which produces red light and is used to treat skin cancers.

1.9 Nitrogen Lasers

The nitrogen laser produces 337-nm radiation through the combined electronic-vibrational transition of molecular nitrogen. Similar to the operating principle of the excimer laser, the nitrogen laser uses a pressurized chamber (0.02-1+ atmosphere) and a pulsed electric discharge which produces relatively high gain, short laser pulses at high peak powers. This excitation mechanism is efficient.

The limitations of the nitrogen laser with respect to power stem from the lower lasing level which has a 10-μsec lifetime. This extinguishes lasing action and population inversion very quickly, resulting in low laser efficiency ($< 0.15\%$). Typical pulse energies are 10 mJ, and pulse lengths are only a few nanoseconds (1).

1.10 Helium-Neon Lasers

The helium-neon (He-Ne) laser was the first gas laser to be demonstrated. Similar to the excimer laser, the He-Ne derives its excitation energy from an electric discharge with an ignition voltage of about 10,000 W which is used to excite helium atoms in a mixture (5:1) of helium and neon gases. After the gas is broken down and can carry an electric current, 2000 V is used to sustain the several milliamperes current during operation.

Excited helium atoms transfer their energy, through collisions in the gas, to neon atoms, which then give off light energy (or energetic photons) as they drop to a lower laser energy level. Helium atoms then reexcite the neon atoms to the upper laser energy level from the ground state.

The light emitted is red in color, characteristic of neon, at a wavelength of 632.8 nm. A weaker He-Ne is used at the weaker 543-nm green line. While the He-Ne does not provide ultraviolet light, it is included here because it plays a key role as an optical alignment tool for UV beam delivery systems. Optical alignment in the ultraviolet is especially critical. The narrow, continuous beam of a He-Ne can operate reliably for thousands of hours at very low (one hundredth of a watt) power. The He-Ne laser is simple, rugged, and inexpensive. The highly visible red beam makes this laser very popular for applications in measuring, aligning (optics, surveying), scanning, and many other uses (1).

1.11 The Alexandrite Laser

The Alexandrite is a tunable solid state vibronic laser, which can supply ultraviolet energy for low dose applications, and is used mainly as a research tool. The Alexandrite laser is named for $BeAl_2O_4$, a mineral in which chromium is added to allow light production. Commercial applications exist mainly in the medical field (tattoo removal) and in scientific instrument markets, but most Alexandrite lasers are in R and D labs.

The Alexandrite is similar to the Nd:YAG in having approximately the same rod size and operating in a continuous or pulsed mode. However, the Alexandrite operates with lower gain, making cavity design more critical. The typical average power of the Alexandrite is in the 80- to 100-W range, placing it in the category of "powerful" lasers. Energy levels achieved with the Alexandrite are similar to the ruby laser, but not as efficient.

Output Power and Wavelength: The gain of the Alexandrite, like other tunable vibronic lasers, is a function of wavelength. The peak is near the center of the laser band, and both gain and output power drop off at shorter (and longer) wavelengths.

The Alexandrite laser emits at wavelengths in the 700- to 825-nm range. The range of tunability is something less than this, depending on the conditions used. Commercial Alexandrite lasers operate untuned at a peak

1.11. THE ALEXANDRITE LASER

emission of 755 nm, with 0.2- to 0.8-J pulses and repetition rates up to 50 Hz. Average power can therefore be several watts. Continuous wave output has exceeded 75 W using a mercury lamp as a pump. Average powers with pulse versions have exceeded 120 W using long pulses.

The temporal output parameters of the Alexandrite are either in Q-switched mode, with 25- 225-nm pulses, or in long pulsed mode with 100- to 400-μsec pulses; continuous operation is also used, mainly in the laboratory.

Spectral Bandwidth: The spectral bandwidth of tunable vibronic lasers is a function of the cavity optics, and since most lasers of this type are research systems, line widths vary considerably. An average figure for spectral line width would be 0.018-0.06 nm (tunable); untuned versions would be about 3 nm.

Operating Efficiency: The basic operating efficiency (electrical-to-optical) of the Alexandrite is in the 0.075-1.25% range, improving with higher quality crystal material. The slope efficiency (ratio of the increase in output power to the increase in input electrical power above the lasing threshold) is about 7.4%. The slope efficiency of a CW Alexandrite has been reported to be 50% (conversion of optical pump energy, not overall efficiency) (7).

Facilities, Safety, and Consumables: The Alexandrite laser is generally water cooled, operating off a 220-V line with 20-A circuits. The only consumables other than water are the lamps used for pumping and occasionally a damaged crystal. Special water chillers are used to cool the laser.

Safety precautions are the same as for any laser with a wavelength capable of reaching the retina of the eye. Retinal sensitivity drops off logarithmically at wavelengths beyond 650 nm or just before the end of the visible spectrum; the human eye can detect wavelengths slightly beyond 755 nm. The danger at these long wavelengths is having a laser with an intense beam at 800 nm, which to the human eye is only a faint beam. Near-infrared radiation cannot only reach the retina, but can cause permanent damage at relatively low energies.

The long wavelength IR energy from lasers, if not reaching the retina, can and will damage the outer portions of the eye by burning the tissue, just as UV does. These are all short-term effects; long-term damage caused by IR energy is not well documented at this writing. Shorter wavelengths (ultraviolet) require standard UV-absorbing eye protection and face shields.

A good safety precaution for eye protection is to first measure and/or determine the frequencies emitted by the laser in question. Then, using published commercial specifications from companies selling laser goggles, select the model that blocks that part of the spectrum being emitted. Vibronic lasers are not widely used, and standard laser goggles may not block emission lines from these lasers.

The other precaution for vibronic lasers is the damage from high intensity, Q-switched versions. These pulses will immediately damage the outer eye (cornea) and possibly penetrate and damage the inner portion of the eyeball, including the retina.

References

1. Hecht, J., "Understanding Lasers," p. 24, Howard and Sans Co., 1988.

2. Elliott, D., "Integrated Circuit Fabrication Technology," 2nd Edition, McGraw Hill, New York, NY, 1993.

3. Ganz, J., "Laser Technology: New Promising Applications," Cooper Vision, Santa Clara, CA, 1995.

4. Bachur, J., "Sources for the Production and Control of Deep Ultraviolet Radiation," Hybrid Technology Group, San Jose, CA.

5. Coleman, L., "Lasers and Applications," Publication by Lawrence Livermore Labs, Livermore, CA, 1992.

6. Lizotte, T., Ohar, O., "UV Nd:YAG Laser Technology," Lasers and Optronics, Sept., 1994.

7. Hecht, J. "Laser and Applications," September 1984.

Glossary of Terms

Actinic Relating to or exhibiting actinism. The property of radiant energy, especially in the visible and ultraviolet spectral regions by which chemical changes are produced.

Black Light Radiation from the ultraviolet or other region of the electromagnetic spectrum where radiation is visible.

Fluorescence A form or type of luminescence with a very short lifetime.

Foot Candle Unit of illumination equal to luminous flux density of one lumen per square foot.

1.11. THE ALEXANDRITE LASER

Frequency The reciprocal of time required by a light wave to oscillate through a full cycle.

Harmonic Generation A nonlinear optical process in which a laser beam is converted to one-half (second harmonic), one-third (third harmonic), one-quarter (fourth harmonic), or some fractional multiple of its frequency, usually by processing the beam through a crystal.

Index of Refraction The speed of light in a vacuum divided by the speed of light in a material.

Interstitial Presence of an ion in a lattice point normally unoccupied.

Lepton Particles with one-half spin which are not subject to the strong nuclear force, but are subject to the weak force and, if charged, to the electromagnetic force (muons, electrons, tautons, neutrinos, and their antipartic).

Light Visible electromagnetic radiation; may include near-infrared and near-ultraviolet portions of the electromagnetic spectrum.

Peak Power The maximum instantaneous energy level in a laser pulse.

Phase Matching The precision (few parts per million) matching of the refractive in discs of the fundamental laser beam and the (its) orthogonally polarized harmonic; all harmonic frequencies on the fundamental beam path are phase matched in the forward direction, resettling in a single, harmonic beam.

Phosphorescence A type of luminescence with a characteristic long lifetime.

Photon A quantum of electromagnetic energy of a single mode (a single wavelength, polarization, and direction). A photon is a unit of energy which equals $h\nu$ (h = Planck's constant; ν = the frequency of the propogating wave). The formulas to calculate the momentum of the photon in the direction of propagation = $h\nu/c$, where "c" is equal to the velocity of light; a quantum of light of a precise wavelength with no charge or rest mass and intrinsic spin 1.

Picosecond One trillionth ($10-12$) of a second.

Plane Wave A light wave where surfaces in constant phase form parallel planes.

Polarity, Photomask The negative (dark field) or positive (clear field) orientation of a photomask.

Population Inversion A condition in a laser medium (cavity or gain medium) when the number of atoms at upper energy levels exceeds the number of atoms at lower energy levels as a result of pumping.

Pumping The inputting of energy to a laser medium to produce population immersion; pumping increases the number of atoms at upper energy levels until they become predominant (population inversion); a necessary condition for lasing.

Radiance Radiant flux per unit solid angle (intensity) per unit area of an element on an extended source or reflecting a surface in a given direction.

Radiant Flux Instantaneous power in watts.

Rays The path of light indicated by a line.

Reflectance The fraction of the total luminant flux incident upon a surface that is reflected and which varies according to the wavelength distribution of the incident light. Ratio of light reflected to light incident on a surface.

Reflection Method by which incident light leaves a surface or medium from the incident side until no change in wavelength.

Refraction Light bending that occurs when a ray enters a material of a different refractive index than the one it was passing through, and is directionally changed as a function of the refractive index differences of the two mediums.

Refractive Index The figure that represents the difference between the speed of light in a vacuum and its speed in the material being tested.

Speckle Granular-appearing laser radiation (noise) caused by the superposition of scatter light having random phase difference.

1.11. THE ALEXANDRITE LASER

Spontaneous Emission The emission of random incoherent light without outside stimulation caused by an atom or molecule moving from a high energy state to a lower one.

Standing Waves Periodic variations in the incident and reflected exposure energy of a signal causing constructive and destructive light interference patterns which are often replicated in the material being exposed.

Stimulated Emission The emission of photons from excited species promoted by the interaction of other photons.

Superradiance The spontaneous emission of energy from a group of atoms with collective phase coherence.

Transverse Mode Modes across the width of a laser resonator.

Ultraphotic Rays Rays of energy that lie beyond the visible portion of the electromagnetic spectrum.

Ultraviolet Wavelengths in the electromagnetic spectrum. The region of the electromagnetic spectrum that begins at the end of the "violet" portion of the visible spectrum (400 nm), and extends down to where the x-ray region begins (1 nm).

Ultraviolet Lens A lens fabricated with a material that is transparent to UV radiation. UV lens materials include high purity fused silica, quartz, calcium fluoride, lithium fluoride, and combinations of materials for compound or multielement lenses, such as a quartz-fluoride achromat.

Ultraviolet Photoemission Spectroscopy A method of measuring the energy spectrum of electrons which are emitted when a material absorbs UV radiation. The spectrum will indicate the characteristic ionization energies of the molecules absorbing UV light, information useful in understanding the nature of UV interaction and the physical and chemical nature of the material being studied.

UV Lamp A lamp that emits a sizable quantity of ultraviolet radiation. Examples are mercury arc and xenon or deuterum lamps which are enclosed by a quartz envelope. The lamps may contain "dupont" gases to increase the spectral output in the UV region.

UV Laser A laser that emits radiation in the ultraviolet region, such as an excimer laser, quadrupled Nd:YAG laser, or He:Ne laser.

Vacuum Ultraviolet The region of the ultraviolet spectrum between 200 nm and the shortest end of the UV spectrum, about 10 nm.

Vacuum Ultraviolet Spectroscopy A spectral analysis technique that uses UV wavelengths in the 100- to 300-nm range for absorption and emission-based analysis.

Vacuum Ultraviolet (VUV) Radiation Radiation in the spectrum of 100-300 nm, generally requiring the use of a vacuum environment to prevent the absorption of the VUV energy by gas (air) molecules.

Visible Light Electromagnetic radiation visible to the human eye with wavelengths from 700 to 40 nm.

Visible Spectrum The region of the electromagnetic spectrum that can be perceived by the eye or that the human retina is sensitive to, or 400- to 750-nm wavelengths.

Wavelength The distance between phase maxima or wave peaks in a light beam. Ultraviolet wavelengths vary from 400 nm down to 10 nm; a single complete cycle of oscillation of light.

Wavefront A surface of constant phase in a propagating wave.

Wave Number The number of 1/2 wavelengths in a centimeter (cm^{-1}).

Reading Bibliography

1. Chen, K.M., Tzeng, M.H., Kuo, C.N., Pei, C.C., "Characterization of Luminous Beams from Laser Ablation by Imaging Techniques," Journal of Chemical Physics Letters, Vol. 232, No. 5-6, p. 576-80, Jan. 1995.

2. Le Blanc, S.P., Qi, Z., Sauerbrey, R., "Femtosecond Vacuum-Ultraviolet Pulse Measurement by Field-Ionization Dynamics," Optics Letters, Vol. 20, No. 3, p. 312-4, Feb. 1995.

3. Witanachchi, S., Ahmed, K., Sakthivel, P., Mukherjee, P., "Dual-Laser Ablation for Particulate-Free Film Growth," Applied Physics Letters, Vol. 66, No. 12, p. 1469-71, March 1995.

1.11. THE ALEXANDRITE LASER

4. Costela, A., Figuera, J.M., Florido, F., Garcia-Moreno, I., Collar, E.P., Sastre, R., "Ablation of Poly (Methyl Methacrylate) and Poly (2-Hydroxyethyl Methacrylate) by 308, 222 and 193 nm Excimer Laser Radiation," Applied Physics A (Materials Science Processing), Vol. A60, No. 3, p. 261-70, March 1995.

5. Burimov, V.I., Zherikhin, A.N., Popkov, V.L., "Investigation of the Populations of Excited States of Barium Atoms in a Laser Plasma," Quantum Electronics, Vol. 25, No. 2, p. 143-5, Feb 1995.

6. Whybrew, A., Webb, C.E., "Nitrogen Dioxide: A Variable Attenuator for KrF Laser Radiation," Measurement Science and Technology, Vol. 6, No. 4, p. 431-4, April 1995.

7. Bisson, S.E., "Parametric Study of an Excimer-Pumped, Nitrogen Raman Shifter for Lidar Applications," Applied Optics, Vol. 34, No. 18, p. 3406-12, June 1995.

8. Pnachenko, A.N., Taransenko, V.F., "Maximum Performance of Discharge-Pumped Exciplex Laser at Lambda=222 nm," IEEE Journal of Quantum Electronics, Vol. 31, No. 7, p. 1231-6, July 1995.

9. Mossavi, K., Fricke, L., Liu, P., Wellegehausen, B., "Generation of High-Power Subpicosecond Pulses at 155 nm," Optics Letters, Vol. 20, No. 12, p. 1403-5, June 1995.

10. Faris, G.W., Dyer, M.J., "Two-Photon Excitation of Neon at 133 nm," Optics Letters, Vol. 18, No. 5, p. 382-4, March 1993.

11. Fontaine, B., Sentis, M., "New VUV Primary Sources by Collisional Pulses," Annales de Physique, Vol. 17 colloq, No. 1, p. 219-27, June 1992.

12. Hecht, J., "Excimer lasers produce powerful ultraviolet pulses," Laser Focus World, Vol.28, No. 6, P. 63-4, 66, 68, 70-2, 1992.

13. Taira, Y., "High-power continuous-wave ultraviolet generation by frequency doubling of an argon laser," Japanese Journal of Applied Physics, Part 2, Vol. 31, No. 6A, P. L682-4, 1992.

14. Umstadter, D., Yu, L.H., Johnson, E., Li, D., "Ultrashort ultraviolet free-electron lasers," Journal of X-Ray Science and Technology, Vol. 4, No. 4, P. 263-74, 1995.

15. "Visible and UV Lasers," SPIE, Vol. 2115, 1995.

16. Hammer, J.W., Petkau, K., Griegel, T., Langhoff, H. Mantel, M., "New possible laser media in the UV and XUV in high current heavy ion beam induced plasmas," Hyperfine Interactions, Vol. 88, No. 4, P. 151-6, 1994.

17. Lister, J.M., Divall, E.J., Downes, S.W., Edwards, C.B., Hirst, G.J., Hooker, C.J., Key, M.H., Ross, I. N., Shaw, M.J., Toner, W.T., "Sprite- a very high brightness, ultraviolet laser systems," Journal of Modern Optics, Vol. 41, No. 6, P. 1203-15, 1994.

18. Hooker, S.M., Webb, C.E., "Progress in vacuum ultraviolet lasers," Progress in Quantum Electronics, Vol. 18, No. 3, P. 227-74, 1994.

19. Szatmari, S., "High-brightness ultraviolet excimer lasers," Applied Physics B (Lasers and Optics), Vol. B58, No. 3, P. 211-23, 1994.

20. Okada, F., Togawa, S., Ohta, K., Koda, S., "Solid-state ultraviolet tunable laser: a $Ce^3\pm$ doped $LiYF_4$ crystal," Journal of Applied Physics, Vol. 75, No. 1, P. 49-53, 1994.

21. Skakun, V.S., Tarasenko, V.F., Fomin, E.A., Kuznetsov, A.A., "Ultraviolet and vacuum ultraviolet excimer lamps pumped by a barrier discharge," Zhurnal Tekhnicheskoi Fiziki, Vol. 64, No. 10, 1995.

22. Glownia, J.H., Gnass, D.R., Sorokin, P.P., "Recent developments in VUV femtosecond pulse generation: Proposal for a higher efficiency F_2 (157nm) laser," SPIE Vol. 2116, P. 208-18, 1994.

23. Kane, D.J., Taylor, A.J., Trebino, R., DeLong, K.W., "Single-shot measurement of the intensity and phase of a femtosecond UV laser pulse with frequency-resolved optical grating," Optics Letters, Vol. 19, No. 14, P. 1061-3, 1994.

24. Kitamura, M., Mitsuka, K., Sato, H., "A practical high-power excimer lamp excited by a microwave discharge," Applied Surface Science, Vol., 79-80, P. 507-13, 1994.

25. Seliskar, C.J., Landis, D.A., Kauffman, J.M., Aziz, M.A., Steppel, R.N., Kelley, C.J., Qin, Y., Ghiorghis, A., "Characterization of new excimer pumped UV laser dyes." III Laser Chemistry, Vol. 13, No. 1, P. 19-28, 1994.

Chapter 2

Ablation

2.1 The Ablation Phenomenon

The interaction of UV laser light with solid matter is an extremely complex phenomenon, and the precise mechanisms involved in UV laser ablation are still being studied and debated.

Ablation is not a single event, but is a complex interaction of many overlapping and separate events, as will be described below. Some of the events include absorption, electronic excitation, rapid heating, plasma generation, photolysis, sputtering, melting, expansion, ejection, decomposition, oxidation, vaporization, sublimation, condensation, and solidification.

In practice, the absorption of UV photons by organic and inorganic molecules will give rise to electronic excitation. The subsequent reaction pathways following initial absorption can be varied and are complex, dependent on the laser wavelength, laser fluence, the exposure medium (gas atmosphere, for example), properties of the irradiated material, and temperature and pressure conditions. A single schematic model of the ablation phenomenon with the inert or reactive gas flow to treat or move ablative debris is shown in Figure 2.1.

When UV energy, in this case from a UV laser, is absorbed by the surface of a solid, the electromagnetic energy of the beam is converted to mechanical, thermal, chemical, and electronic energy, and the resulting ejected ablative debris, in the form of a plume, includes atoms, molecules, ions, photons, electrons, and agglomerations or fragments of the irradiated material.

Several models have been presented which attempt to characterize the electronic, temporal, thermal, and energy gradient aspects of ablation. We

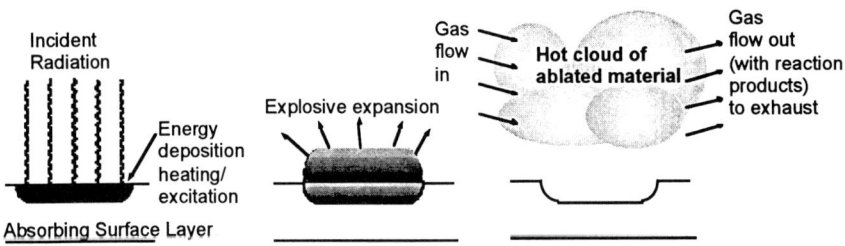

Figure 2.1: Schematic of UV Laser Ablation Events

will present highlights of some of these models in this chapter, along with empirical results of some key ablation experiments.

One area of interest in studying ablation reactions is the ability to control the UV interaction parameters and achieve athermal processing. IR photons, on the other hand, had been observed to cause vibrational and rotational excitation leading to high thermal-content reactions. It is precisely this key difference in the interaction of UV and IR photons which led to the use of UV light as a microstructuring tool for solids, as well as a "driver" for a variety of reactions where control of etch depth or thickness of deposition of a layer is critical, such as in semiconductor processing.

Ultraviolet light, by reproducing its shape in a solid via the ablation phenomenon, is an excellent tool for microlithography, microsurgery, micromachining, and similar applications involving delicate, controlled exposure, cutting, or removal of material.

2.2 Background and History of Ablation

The emergence of the optical laser in the 1960s is the recognized beginning of laser technology. The first ablation experiments were simply observations of the interactions of laser light and solids, and resulting phenomenon. The work of Breech and Cross in 1962 involved the use of a ruby laser as an energy source to excite and vaporize a solid for the purpose of elemental analysis. Later, this was called "laser microprobe emission spectroscopy." While spectroscopy relies on the analysis of photons to identify elemental composition, the ablation phenomon with the ejection of gas, solids, and

2.3 Early Discoveries and Uses of Ablation

Early work identified a practical use for the phenomenon of ablation, and commercial tools were made for experiments in medical, geological, and metallurgical fields. The main commercial applications were for laser microprobe emission and mass spectrometry. Other practical uses that emerged for ablation were high power lasers for cutting, welding, marking/scribing of nomenclature, and micromachining of holes. An early text describing high power laser experiments and ablation reactions is "Effects of High Power Laser Radiation" (2).

Early lasers were not powerful, and the first experiments were made with long wavelength, red lasers. In the mid-1960s, visible, near-UV, and UV tests were run, and more powerful sources were deployed, leading to the discovery of multiphoton effects. An early paper described hypersonic and ultrasonic oscillations during ablation to increase the ejection of material in the plume (3). Another experimental study (4), the first on semiconductors, recorded the number of positive ions (10^8) and electrons (3×10^{16}) from a single laser pulse. They counted the number of atoms ejected by ablation (2×10^{17}) and the size and depth of the ablated area. This is considered the first real experiment with useful data on ablation, published in 1963.

Characterizing some of the key aspects of the ablation phenomenon was done by J.F. Ready and J.A. Howe, who measured the temporal and spatial profile of a plume, the complete volume of ejected materials (electrons, gas, solids, ions, photons, neutrals, clusters) from an ablation reaction, and measured temperatures of the fluorescence of CN and C molecules.

Some of the observations of their experiments in ablating a carbon sample are as follows:

- Plume occurrence 120 nsec from beginning of pulse

- Plume velocity 2×10^6 cm/sec

- Plume duration 5-9 μsec

- Vibrational temperature 10,000-20,000K

- Rotational (from CO_2 and CN molecules) temperature 4500K

Many subsequent researchers continued this early work by ablating a variety of target materials, mapping their temperatures, velocities, and kinetic energies. Further analysis of ablation plumes was made, including mass and charge dependence; studies of postexcitation and postionization neutrals were made to further characterize the plume dynamics. The first models of laser ablation were written, and a more complete understanding of the entire kinetics and physics of ablation was emerging.

The practical application that emerged from laser ablation of solids was laser microprobe emission spectroscopy. Lasers improved in reliability, and the first commercial tool based on this technique was launched in 1978.

In the 1980s and early 1990s, the emergence of the excimer laser brought increased activity to ultraviolet laser ablation; Nd:YAG and CO_2, the primary energy sources for ablation studies, were joined by high energy tunable pulsed UV lasers and picosecond lasers.

The combination of new and more reliable laser sources and an array of advanced analytical and diagnostic equipment resulted in the ability to fully map the behavior of laser ablation reactions and to produce models to describe all the phenomenon occurring in the complexity of reactions. Materials research, once the primary application for laser ablation, started to give way to actual uses. Microsurgery, microlithography, and micromachining became the first major practical end uses for UV laser technology.

Laser-material interaction studies and applications have now become widespread, and commercial uses of laser ablation are now emerging in several market areas.

2.4 Ablation Mechanisms

Initial theories of ablation were based on rapid thermalization. Figure 2.2 illustrates the effects of the thermal role in ablation, showing ablated glass or silicon dioxide, about 12000 Å thick, ablated to expose away a metal conductor line in a failure analysis application. Presently, it is not completely clear which mechanisms dominate, but there is a mounting body of evidence suggesting that electronic effects may be responsible for most of the processes occurring in the laser ablation of solids (nonmetals). Most of the experimental evidence presented here is based on the ablation of materials such as polyimide, photoresist, and semiconductor films such as silicon dioxide.

2.4. ABLATION MECHANISMS

Figure 2.2: SiO$_2$ Layer Ablated over an Aluminum Conductor/SEM at 1100x (Courtesy of IMS)

2.4.1 Molecular Bond Breaking

The typical bonding and extinction energy for many molecules is in the 3- to 15-eV range, which corresponds roughly to the range of photon energy in the ultraviolet. UV photons, from 157 to 308 nm, provide sufficient energy for breaking bonds in a wide variety of materials. Chapter 6 on UV laser cleaning covers bond breaking in detail.

Color center formation may be induced by an initial pulse in a long pulse train. Formation of a color center will result in a change in the absorption properties of the irradiated material as a single pulse moves through the solid.

Related electronic effects include other forms of laser conditioning of both surface and bulk material which have an effect in subsequent ablation processes, such as a change in the rate of ablation or preferential ejection of conditioned atoms or clusters.

2.4.2 Absorption, Heating, and Explosive Expansion

Photon absorption on the surface of a solid material produces several physiochemical reactions, all of which are part of the ablation process.

Intense absorption of UV photons, as from a 248-nm excimer laser pulse, causes heating of the surface of the absorbing material up to several hundred degrees centigrade. The heat is generated within the time frame of the UV pulse duration, or about 10-20 nsec. Since this is too short a time for thermal conduction and subsequent dissipation of the energy into the bulk of the solid, the heating effect is localized to a very thin (1000-2000 Å) layer of material.

If the absorption intensity exceeds a characteristic threshold point (the ablation threshold) for the irradiated material, the result is a volume explosion, a catastrophic thermal expansion away from the surface of gases, ions, molecules, fragments and clusters of solid debris, and photons. The products of photodecomposition have a larger specific volume than the original unexposed material, giving rise to the explosive expansion and mass transport away from the surface of the irradiated solid. The motion of the ejected material is near normal to the surface, and the mean perpendicular velocity of 2200 m/sec is determined by recoil analysis (5).

As an example, liquid methracylate has a specific volume 30% larger than polymethylmethracylate (PMMA). While the explosive expansion of material carries away a thin surface layer of the solid, the underlying material immediately adjacent to it is largely unaffected, such as the sidewall and bottom surfaces. For this reason, UV laser ablation is typically characterized in terms of its effect on the remaining unexposed solid. The use of UV laser energy has obvious applications for structuring heat sensitive materials.

2.4.3 Fluorescence and Photoionization

Fluorescence is typically defined as: Intense UV exposure of the surface of a solid will raise molecules and atoms to excited states, followed by relaxation and emission of photons of different wavelengths, or fluorescence.

The high levels of excitation reached in these reactions also cause ejection of electrons, or photoionization.

2.4.4 Ablation Plume and By-products

The ejected material, referred to as a "plume," is a vibrationally and electronically hot mixture with a wide span of molecular weight. This "ablative"

2.4. ABLATION MECHANISMS

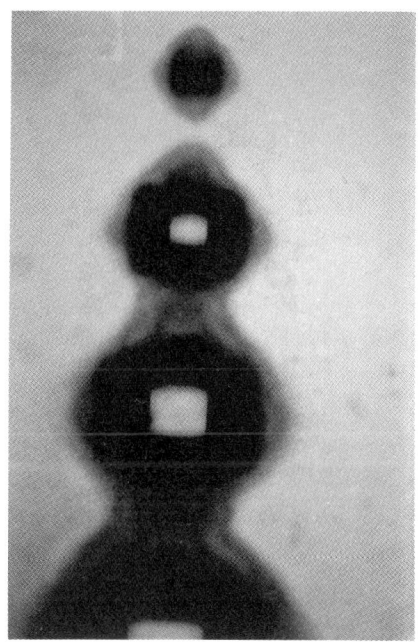

Figure 2.3: Ablation Plumes from Polymer Machining

40   CHAPTER 2. ABLATION

debris reaches very high velocities, then recondenses into a thin sooty carboniferous deposit, illustrated in Figure 2.3.

During ablation of organic materials, by-products include carbon monoxide, carbon dioxide, water, and a variety of compounds with varying levels of carbon (C_2-C_{12}). In one experiment, a 193-nm excimer laser was used to ablate polyethylene terephthalate films in a vacuum. Analysis was performed using GC/MS. The by-products of the reaction were analyzed to help understand ablation mechanisms.

2.4.5 Spectral Absorption

A primary requirement for an ablative photodecomposition reaction is strong absorption of the UV photons by the solid. Therefore, the first information required to properly estimate UV photon interaction properties is a spectral transmission profile. Figure 2.4 shows the transmission versus wavelength data for fused silica (30). This indicates that good absorption of UV photons, especially below 190 nm, will occur.

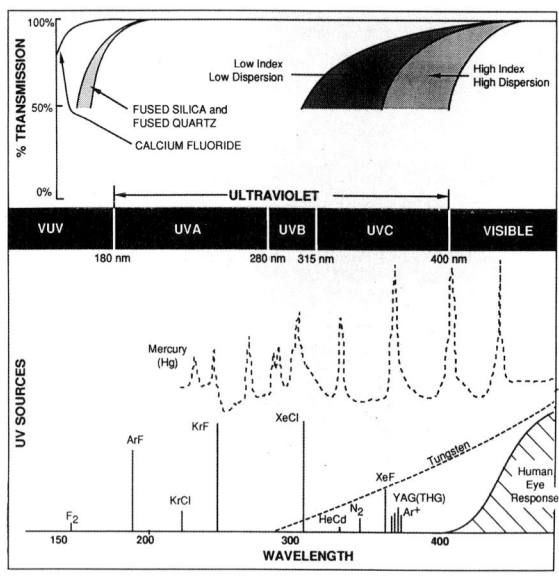

Figure 2.4: Spectral Characteristics of Optical Materials

2.4. ABLATION MECHANISMS

The ability to select specific wavelengths of light with the excimer laser permits optimization of light coupling into the substrate. By choosing a wavelength that most closely matches the spectral absorption of the substrate, increases will occur in the efficiency of the cutting reaction as well as the resolution and edge sharpness. After determining the spectral absorption data of the material to be cut, the appropriate wavelength can be chosen. Optical coupling efficiency is optimized when the spectral output of the laser energy source matches the absorption coefficient of the substrate.

2.4.6 Absorption Calculation

The actual UV photon absorption mechanism through a solid follows Beers law:

$$I/e = I/o \times 10^{-\epsilon\alpha} \tag{2.1}$$

where I/o is laser beam intensity before transmission, I/t is laser beam intensity after transmission, ϵ is material thickness, and α is absorptivity of the solid.

In strongly absorbing materials, such as photoresist films, photon absorption causes electronic (vibrational and rotational) excitation within the molecule of the material being exposed to the laser beam. Sufficient incident energy will overcome the bonding energy of the molecule, rupturing the molecule into atomic, molecular, and larger fragments.

2.4.7 Ablation Threshold

Essentially a dry photoetching technique, ablative photodecomposition relies upon the geometry of the light source (the excimer laser beam) to define the geometry created in a given solid.

The level of energy required to cause the breaking apart of solid materials by UV photon absorption is termed the *ablation threshold*. The ablation thresholds for a variety of materials at 308 nm are shown in Table 2.1 (32). The 308-nm wavelength is used in excimer laser micromachining.

UV absorption mechanisms in solids, and the resultant decomposition pathways, are complex. A good example to explain these mechanisms is provided by R. Srinivasan (24). In microelectronics processing, the synthetic organic polymer PMMA is, like most polymers, an organic solid composed principally of carbon, hydrogen, oxygen, and nitrogen. This polymer is composed of many small units called *monomers*, each up to about 50 atoms

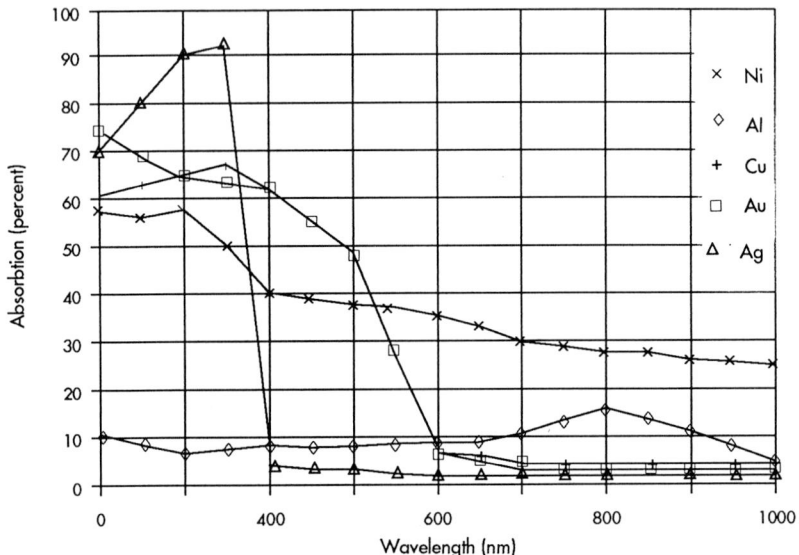

Table 2.1: Ablation Thresholds for Microelectronic Thin Film Materials

long, which are repeated throughout the entire polymer to form a polymer chain. The structure of PMMA is shown in Figure 2.5.

Atomic bonding in a polymer is relatively strong (60-150 kcal/mol); however, molecular bonds are relatively weak, on the order of 10 kcal/mol or less (10). An energy diagram is used to illustrate UV photon absorption by such a single bond (A-B) in a molecule, shown in Figure 2.6.

The bond ground state is shown in the lowest curve. The vibrational levels within a specific electronic level are indicated by the short horizontal lines. The pathway from ground state to the upper excited level is indicated by the #1 vertical arrow. This reflects the absorption of a UV photon and is believed to occur so rapidly that it proceeds without nuclear motion. The upper state is metastable, containing sufficient energy to allow the atoms to dissociate at the next single vibration. The horizontal arrow (#2) indicates internal conversion, wherein the excited species cross to the ground state energy surface with sufficient energy for dissociation.

In addition to internal conversion, many other possible pathways (mostly competing reactions) exist following the relatively simple process of excitation. These include fluorescence (the reverse of excitation), vibrational deactivation, collisional quenching, and intersystem electronic state changes.

A complex set of reactions consisting of both photochemical and pho-

2.4. ABLATION MECHANISMS

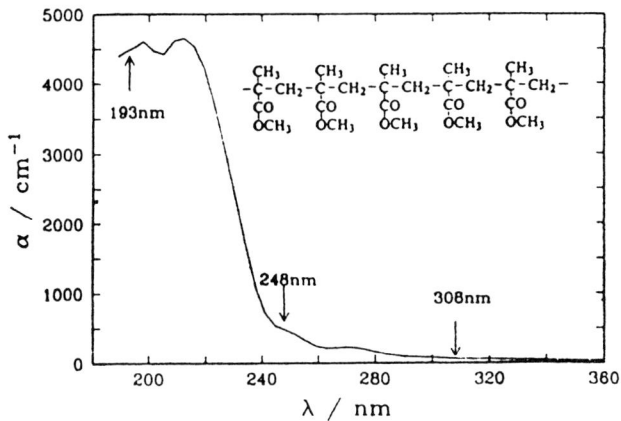

Figure 2.5: PMMA Structure and Spectral Absorption

tothermal reactions occur. The reader is referred to the text *Ablation* from Springer-Verlag for detailed descriptions of the photophysics at work in these reactions.

In short, the UV absorption mechanism is one where the UV photons excite bonding electrons and specific chromophores (atomic groups) are responsible for specific absorptions. In the regime above 190 nm, UV photons cause a valence electron to be promoted from a bonding to an anti-bonding orbital. Increasing the photon energy simply increases the number of valence excitations, resulting in more complex photoreactions, thereby increasing the difficulty of correlating a specific transition within a polyatomic molecule. M.B. Rubin has explored this subject in a study of the higher excited states of polyatomic molecules (24).

The various transition reactions cited earlier for PMMA, polyimide, and similar polymers can occur singly or in some combination simultaneously. The energy level diagram shows the two pathways where UV photon absorption leads to ablation reactions. The first path (#1 in Figure 2.6) relies on photon energy alone, without thermal effects, and is termed *photochemical ablation*. The second path (#2 in Figure 2.6), where decomposition occurs in the vibrationally excited ground state, is considered *photothermal ablation* due to the role of heat in the decomposition of the bonds. The first

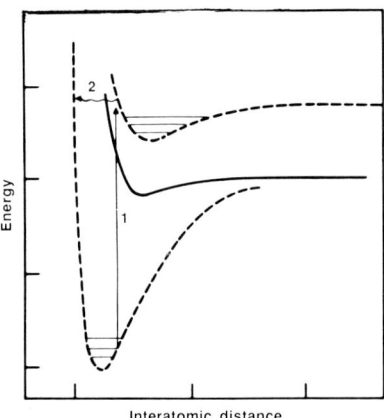

Figure 2.6: Energy Level Diagram for a Hypothetical Molecule A-B

path allows for material decomposition in the upper level excited electronic state.

Pure photochemical ablation offers the advantage of higher resolution and sharper edge acuity in cutting due to the absence of thermal distortion and other thermal effects, such as cracking, crazing, charring, carbonizing, graphitizing, or melting. These effects are typically present when UV photons are used as a source "work heat."

2.4.8 Ablation Parameters

The effects on the substrate during ablation are dependent on both laser parameters and material parameters. Laser parameters include wavelength, reprate, pulse length, pulse energy, and coherence. Material parameters include substrate type and thickness, spectral absorption, substrate reflectivity, and density.

Laser pulse lengths typically range from 10 to 30 nsec. Ablation (photochemical or photothermal) occurs within this length of time, and the diffusion length of a single pulse is about 700 Å in a typical polymer (varies according to material type). Therefore, thermal damage to the underlying substrate is unlikely. In most cases, the amount of energy absorbed during UV laser

2.4. ABLATION MECHANISMS

irradiation of highly absorbing materials is nearly totally consumed by the irradiated material. Any energy beyond that required to initiate ablation will be expulsed in the plume of ejected debris and gas by-products of the reaction.

A mathematical expression to relate etch depth per pulse to laser fluence, fluence threshold, and absorptivity at a specific wavelength has been given by several researchers (24,25) and relates Beer's law to the UV photon absorption mechanisms. The phenomenon of ablative photodecomposition is highly time and energy dependent. It is also very complex, since reactions can contain a wide range of both photothermal and photochemical mechanisms.

At the onset of intense UV photon absorption by a solid, energy is deposited so rapidly that the material undergoes extreme expansion. Pressure builds up at the atomic level, giving rise to a photoacoustic pulse that exits the material ahead of the solid and gaseous ablation by-products. A photo sequence of this shock wave (26), emanating from an irradiated polymer, is shown in Figure 2.7

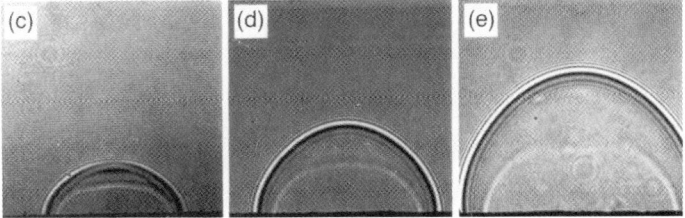

Figure 2.7: Shock Wave Sequence from 193nm Photoablation

Note the relatively long delay before the solid material begins to exit from the sample following the exit of the stress wave. The photoacoustic wave precedes the actual "plume" of ablative debris by a time considerably longer than originally predicted by early photoablation theory.

The interaction of UV laser pulses with solids is further complicated by the fact that the leading edge of a pulse in a solid initiates reactions even before the entire pulse is absorbed. As the pulse is gradually and fully absorbed, a continuum of energy *states* is left in the absorbing material. The

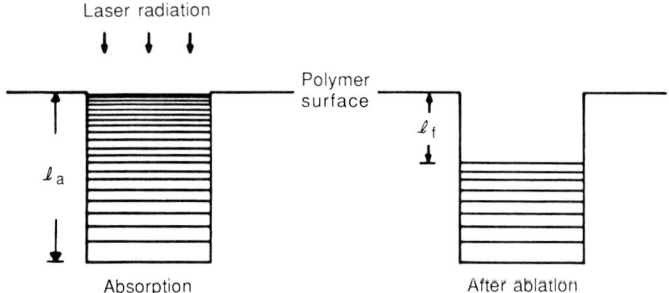

Figure 2.8: Interaction of a Laser Pulse in a Polymer

nature of this continuum, as an uninterrupted, ordered sequence of change in the material, is determined precisely by the quanta of photons actually deposited throughout the layer of material. In the z dimension, this quanta changes as a function of energy losses due to absorption, reflection, and dissipation.

According to Beer's law, this would result in a continuum ranging from partially excited to fully ablated material, where the material receiving a quanta of photons equal or in excess of the ablation threshold is ablated; areas receiving an amount of photons below the ablation threshold undergo simple radiational heating (23). The first laser pulse deposited in a given material is therefore unique setting up an energy state within the absorbing material.

The delay between the time the UV photons are absorbed and actual ablation is about 7 nsec, within the half-width (FWHM-full width, half maximum) of the pulse. This temporal delay appears to be somewhat independent of the material type based on experiments by R. Srinivasan, using PMMA, polyimide, and rabbit cornea, all of which behaved similarly in this respect (24). A cross-sectional schematic view of the interaction of a single laser pulse in a polymer is shown in Figure 2.8.

2.5 Practical Characteristics of UV Reactions

In this section we will discuss the useful properties that characterize UV/ material interactions that lead to the specific applications of UV technology. These characteristics are summerized below and each will be discussed in

2.5. PRACTICAL CHARACTERISTICS OF UV REACTIONS

detail.

Characteristics of UV Reactions

- Multiple Wavelengths
- Surface Reactions
- Athermal Processing
- High Photon Energy
- High Resolution and Coherence

The use of energetic beam processing in IC fabrication is not new. Electron beams have been used for many years to produce high resolution reticles (1-10X masks) for microlithography. Ion beams have similarly been used since the early days of VLSI IC fabrication for implanting dopants, replacing the diffusion furnaces. Long wavelength lasers have been used for some time in semiconductor processing. For example, the CW argon laser is used for large-grain crystal growth; Nd:YAG lasers are used for resistor trimming, marking, localized melting, and annealing. Ruby and other long wavelength lasers are used for a variety of metrology applications (stage metering, profilometry, dopant analysis, particle detection) critical to semiconductor processing (11).

The excimer laser, however, has been applied to more individual steps in IC processing than any other single energy source, primarily because of the specific properties that have practical advantage in high density chip fabrication. In this section, we will discuss the specific characteristics of excimer laser material interactions that make them uniquely suited for microelectronic and many other applications.

2.5.1 Multiple Wavelength Capability

Excimer lasers generate energy at several ultraviolet wavelengths, as determined by the selection of gases used in the discharge chamber. The ability to match the wavelength of the laser to the energy absorption properties of the specific semiconductor film allows for maximum processing efficiency and imaging quality. Increasing the energy coupling efficiency from the laser into the specific film reduces the total dose required to perform a specific function.

Efficient optical coupling of energy into a material will reduce the residual heat generated in laser/material interactions. Specific absorbers are added to polymer and other films to increase the interaction efficiency (26). The availability of several wavelengths provides multiple kinetic pathways to bring about a variety of reactions used in laser-assisted deposition and etching.

The lasing specie and corresponding emission wavelength of excimer and related halogen lasers are shown below.

Lasing Specie	Emission Wavelength (nm)
F_2	157
ArF	193
KrCl	222
KrF	248
ClF	284
XeCl	308
I_2	342
XeF	351, 353

Excimer lasers used in production applications are dedicated to a specific wavelength, but a single laser can be used for two and even three individual wavelengths if necessary. It is simpler to convert to a wavelength with the same halogen, such as from 248 nm (krypton fluoride) to 193 nm (argon fluoride). If the conversion is from a "fluorine-based" to a "chlorine-based" gas, additional time for passivation must be allowed. (See Chapter 3 on the operation of an excimer laser.)

Wavelength conversion also requires replacement of the cavity optics, specifically the maximum rear reflector and the front output coupler; in some lasers, broadband reflective coatings are used to allow the use of two wavelengths with the same set of optics. For example, a laser can be used for both 193 and 248 nm, and the laser cavity optics are then coated with a broadband reflective coating that has high reflection across the 193- to 248-nm band. Maximum reflection is obtained by using a single wavelength coating (see Chapter 5 , UV Optics and Coatings).

When converting from one wavelength to another, the gas lines and discharge chamber need to be flushed with an inert gas. After filling with the new gas, power will be lower than normal until the discharge chamber and

2.5. PRACTICAL CHARACTERISTICS OF UV REACTIONS

lines are fully saturated with the new gas mixture. In research labs, there is ample time to flush the lines and repassivate the lines when converting from one wavelength to another; in a production process, a dedicated link is used to avoid the downtime associated with changing gas mixtures/lasing wavelengths.

The wavelength range available from excimer lasers can be extended in several ways. Raman scattering is one of the most efficient, a technique that takes the input beam and transforms it into either a shorter or a longer wavelength. Raman media can be pumped with different excimer lasers to generate wavelengths that cover a wide area of the ultraviolet spectrum (27). Raman scattering can be used to exactly match the laser output spectrum to the absorption spectrum of the material being irradiated.

2.5.2 Surface-Level Reactions

Processing of a thin layer at a time is useful for most of the applications cited in this book. UV photons have the unique property of being absorbed in only the top surface (1-2000°) of a material. Surface-level reactions are required in cleaning, etching, imaging, and many other applications.

Most of the fabrication steps in IC processing involve the deposition, modification, or removal of very thin surface layers of material. A typical layer thickness in microelectronic device processing is between 1000 and 12,000 Å thick (28,29). The excimer laser, being a pulsed laser, can be programmed to interact with specific thicknesses of materials according to the known absorption depth of an individual pulse in a specific substrate material. Thus, as a means to remove, react, or rearrange surface species in precise amounts of thin layers, the excimer laser is ideally suited for microelectronic fabrication.

Highly energetic UV photons from excimer lasers are intensely absorbed by a large number of semiconductor films, including polymers, oxides and glasses, silicides, and metals. This absorption is a surface phenomenon, being confined to a layer typically less than 2500 Å thick (30). Intense absorption of photons at the surface of the semiconductor device keeps heat from reaching the lower levels where it could adversely affect the dopant concentration and change the electrical properties of the device. High absorption at the surface also results in the high utilization of photons. For example, up to 95% of the photons absorbed into polymers break an individual bond (25).

High bond-breaking efficiency allows the laser beam to be used (optically spread) over a relatively large area, permitting increased production throughput. Also, most of the semiconductor processes are surface driven, beginning with crystal growth, and including oxidation, cleaning, ion implantation, annealing, lithography, etching, and deposition. Figure 2.9 shows a diagramatic sequence of UV absorption in a polymer surface (30).

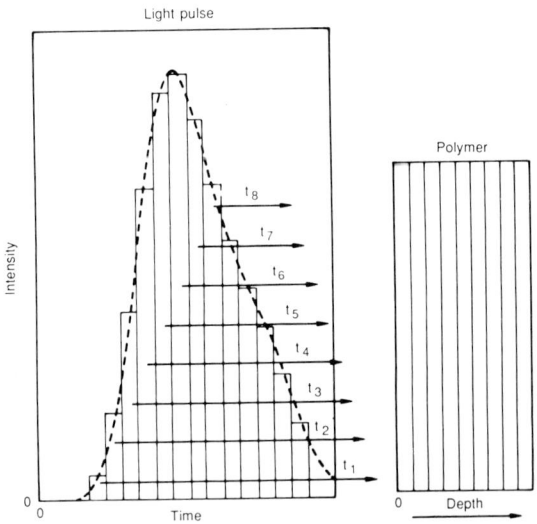

Figure 2.9: UV Energy Absorption in a Polymer Surface

The trend in IC technology is to use thinner layers of material and more individual layers to achieve the speed, density, and overall performance specifications needed in advanced ICs. Confining the reactions to the surface with a UV laser is an ideal application of excimer technology.

2.5.3 Athermal Processing

Traditional methods, such as furnace annealing and infrared baking, lack the degree of control over the reaction depth and thermally distort the substrate. Thermal distortion changes the critical dimensions of the IC pattern and therefore have a limiting effect on electrical specifications.

A unique characteristic of deep-UV energy is its ability to photodissociate low and high density materials without depositing considerable heat in the bulk of the materials. This "athermal" property is especially useful in microelectronics, where delicate thin films need to be structured with as little heat as possible.

2.5. PRACTICAL CHARACTERISTICS OF UV REACTIONS 51

Lasers traditionally used for microelectronic applications have been long wavelength lasers. As a result, they deposit considerable heat, leading to thermal distortion or other thermal side effects. The rapid change in IC geometries to very thin films and submicron geometries dictates the use of low temperature processing. In advanced IC manufacturing, low temperature processing is required in order to avoid distorting the wafer, changing dopant distribution, or causing movement within the various thin film layers.

The interaction of deep-UV, excimer laser energy is limited to a surface region of a few thousand angstrom thickness. Most of the heat escapes outward so that very little is conducted into the bulk of the wafer or mask or other device. Processing with excimer lasers provides a low temperature advantage.

During irridiation of the sample by the excimer laser, vibrational and rotational excitation of the absorbing specie gives rise to considerable heat; the heating causes expansion and breakup of the material. The time scale in which this occurs is subnanosecond, and heat is carried off at a high velocity before it can be absorbed by the sample (30-33). This mechanism is termed "photoablation," or simply "ablation." There is a "photo" and a "thermal" component to the reaction(s) (33). Figure 2.10 shows a schematic diagram of this reaction (24-25).

2.5.4 High Photon Energy

An important property of deep-UV lasers is the relatively high energy of the photons at these wavelengths. Ultraviolet photons intereact strongly with many materials, and as a result can be used to initiate a wide variety of photolytic, chemical, and thermal reactions. The energy per photon increases significantly with decreasing wavelength, as shown below.

	Lasing Wavelength	Energy Per Photon
	Wavelength (nm)	Energy (eV)
F_2	157	7.43
Arf	193	6.42
KrCl	222	5.50
KrF	248	5.00
XeCl	308	4.03
XeF	351	3.53

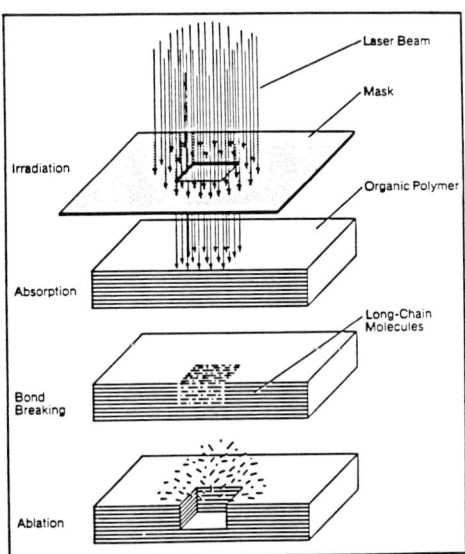

Figure 2.10: Schematic Diagram of an Ablative Photodecomposition Reaction

Other deep-UV wavelengths from nonexcimer laser sources are also used, such as the fourth (266 nm) and fifth (212 nm) harmonics of the YAG laser. Also, the fluorine laser at 157 nm shows the potential of high resolution imaging with high energy photons. Figure 2.9 is an example of high resolution microstructuring of teflon "machined" at 157 nm with a Lambda Physik laser (34).

High UV-photon energy is the unique property that is highly useful in microelectronic processing; it can drive reactions needed for laser-assisted etching, laser-assisted deposition, photolithography, photoablation, annealing, cleaning, and crystal growth.

High photon energy does, however, increase the difficulty in producing UV coatings (for mirrors, windows, and attenuators) and optics (lenses, homogenizers, beam splitters) necessary for laser beam delivery. Research in this area has resulted in increased damage resistance for UV coatings at all of the excimer laser wavelengths, up to several Joules per square centimeter fluence levels. This is sufficient for most microelectronic processing applications. (See Chapter 5, "UV Optics and Coatings.")

Compared to other lasers and doped mercury and rare gas lamps, excimer lasers can deliver significantly higher power. High energy levels are needed to

2.5. PRACTICAL CHARACTERISTICS OF UV REACTIONS

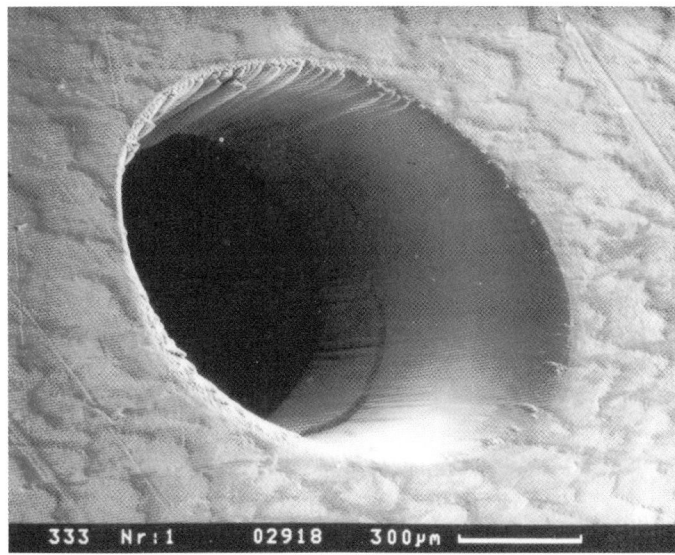

Figure 2.11: Teflon Ablated at 157 nm

achieve the fluence levels required in beam delivery systems, as considerable losses can occur in the optics themselves. Further, sufficient energy must remain to cover the wafer diameter (200 mm typical) for full area processing at acceptable throughput levels.

2.5.5 High Resolution at Short Wavelengths

High resolution is a critical requirement in the manufacture of advanced ICs. The short ultraviolet wavelengths of excimer lasers are being used to image resist films for the microlithography step in IC fabrication The resolution and focal depth advantage over longer conventional lamp-based exposure sources is significant. Excimer lasers at 308, 248, 193, and 157 nm have all demonstrated extremely fine resolution for microlithography, as shown in Figure 2.12. As a result, excimer lasers are now being used to produce advanced semiconductor devices.

The calculation of imaging potential with short wavelengths of light is expressed in the Rayleigh criterion for determining resolution, which shows the relationship of the numerical aperture of the imaging lens (used in excimer laser delivery systems), the imaging wavelength, and a process (K) factor which includes the contrast of the imaging medium.

Figure 2.12: High Resolution (0.3 μm) Photoresist Image Exposed at 248 nm

References

1. E. Arthurs, R. Falster, "Laser Processing of Semiconductors-An Update," Semiconductor International, June, 1982.

2. T. McKee, J. A. Nilson, "Excimer Applications," Laser Focus, June, 1982.

3. D.J. Ehrlich, "Early Applications of Laser Direct Patterning: Direct Writing and Excimer Projection," Solid State Technology, Dec., 1985.

4. D.J. Ehrlich, R.M. Osgood, Jr., T.F. Deutsch, "Photodeposition of Metal Films with Ultraviolet Laser Light," J.Vac.Sci. Technology, 21(1), June, 1982.

5. R.T. Hodgson, W.K. Chou, T.J. McKee, CLEO, 1981.

6. K. Jain, C.G. Wilson, B.J. Lin, IEEE Electron Devices Lett.3,53 (1982); IBM Res. and Dev. 26, 151, (1982).

7. T.F. Deutsch, J.C.C. Fan, G.W. Turner, R.L. Chapman, D.J. Ehrlich, R.M. Osgood, Jr., Appl. Phys. Lett., 38, 144 (1981); D.J. Ehrlich,

2.5. PRACTICAL CHARACTERISTICS OF UV REACTIONS

R.M. Osgood, Jr., T.F. Deutsch, IEEE J. Quantum Electr. 1233, 1980.

8. K. Jain, C.G. Willson, B.J. Lin, "Ultrafast High Resolution Contact Lithography with Excimer Lasers," IBM J. Res. Develop. Vol 26, No.2, March, 1982.

9. D.J. Elliott, "Microlithography-Process Technology for IC Fabrication," McGraw Hill, 1986.

10. S.K. Scarles, G.A. Hart, Appl. Phys. Lett. 27, 243, (1975).

11. T. McGrath, "Appl. of Excimer Lasers in Microelectronics," Solid State Tech., p165, 1983.

12. C. Brau, "Excimer Lasers," ed. C.K. Rhodes (Springer-Verlag, 1979), chap. 4.

13. M. Rokni, J.H. Jacob, "Gas Lasers," ed. E.W.McDaniel, W.L. Nighan, (Academic Press, 1982) chap. 10.

14. T. McKee, J.A. Nilson, "Excimer Applications," Laser Focus, June, 1982.

15. J.J. Ewing, "Excimer Lasers," in 'Lasers Handbook', Vol 3, M.L. Stitch, ed., North Holland, New York (1979).

16. M.J. Shaw, Prog.Quant.Elect., 6,T3, 1979.

17. M.H.R. Hutchinson, Appl.Phys., 21, 45, 1980.

18. I.S. Lakoba, S.I. Yakovlenko, Soviet Journal of Quant. Electron. 10, 389, 1980.

19. H.T. Powell, "Laser Focus," p.57, 1982.

20. J. Frakenthaler, F & F Associates, Private Communication.

21. T.R. Loree, R.C. Sze, D.L. Barker, Appl. Phys. Lett., 31, 37, 1977.

22. S. Wolf, R.N. Tauber, "Silicon Processing for the VLSI Era," Lattice Press, 1986.

23. D. Elliott, "IC Fabrication Technology," Second Edition, McGraw Hill, 1989.

24. R. Srinivasan, "Ablation of Polymers and Biological Tissues by Ultraviolet Lasers," Amer. Assn. for the Adv. of Science, 1986.

25. R. Srinivasan, Bodil Braren, "Ablative Photodecomposition of Polymer Films by Pulsed Far-Ultraviolet (193nm) Radiation: Dependence of Etch Depth on Experimental Conditions," Jol. of Polymer Science, Polymer Chemistry Edition, Vol. 22, 2601-2609, 1984.

26. R. Srinivasan, "Laser Ablation of Organic Polymers: Microscopic Models for Photochemical and Thermal Processes," J. Appl. Phys., 57, (8), 15 April, 1985.

27. Lambda Physik Newsletter, Volume 26, 157nm Lithography.

28. K.G. Ibbs, R.M. Osgood, "Applications of Laser Chemical Techniques to Microelectronics Fabrication," Laser Chemical Processing for Microelectronics, Cambridge University Press, Cambridge, Mass.1989

29. R. Srinivasan, V. Mayne-Banton, Appl. Phys. Lett. 41, 576 (1982); R. Srinivasan and W.J. Leigh, Jour. Am. Chem. Soc. 104, 6784 (1982).

30. M. Dunn, J. Nemecheck, Ted Tienviere, "Making UV Optical Systems Work", Lasers and Optronics, April, 1989.

31. M.B. Rubin "Higher Excited States of Polyatomic Molecules," Academic Press, New York, 1974.

32. G. Shukov, A. Smith, "Micromachining with Excimer Lasers," Lasers and Optronics, Sept. 1988.

33. R. Srinivasan, "Ablation of Polymers and Biological Tissue by Ultraviolet Lasers," Science, 31 Oct., 1986, Vol. 234, pp. 559-565.

34. R. Srinivasan, Journ. Vac. Sci. Technol. B4, 923 (1983);

35. R. Srinivasan and B. Braren, J. Polym. Sci,. Polym. Chem. Vol. 22, 2601-2609, (1984).

Glossary of Terms

Ablation A complex phenomenon occurring when an energy threshold is reached which causes measurable or observable damage and material removal.

2.5. PRACTICAL CHARACTERISTICS OF UV REACTIONS

Absorption The interaction that takes place between the optical electric field of light and the most nearly free electrons of a solid or gas material. In quantum mechanics, the interactions between photons and electrons in the presence of the lattice of nuclei.

Color Centers In laser optical systems, absorbing sites in the lens element induced by repeated exposure to high intensity radiation.

Electron A negatively charged particle revolving around the nucleus of an atom that forms bonds with other atoms. An atomic nucleus without an electron is an ion.

Electronic Excitation Density Volumetric density of electron-hole pairs caused by incident laser exposure on the surface of a solid material.

Electronic Transition A change in an electron's energy level.

Exciton An electron hole pair with an energy lower than the conduction band edge.

Exiton An excited electron-hole pair.

Flux Density The volumetric density of photons per unit time which are able to cause the formation of excited by-products at the surface of the exposed solid.

Laser Ablation A complex process occurring when an energy threshold is reached which causes measurable or observable damage and material removal to the exposed surface; where the rate of removal exceeds one monolayer per pulse with atom, ion, or cluster yields which are superlinear functions of the electronic excitation density.

Laser-Induced Desorption The emission of particles from a surface exposed to high energy laser pulses with no actual damage to the surface. Laser-induced particles emission is often considered a precursor step to laser ablation since it induces changes (rearrangement of structure, defects, etc.) in the surface of the exposed solid which become nucleation sites or intense centers for high energy absorption that lead more readily to ablation than other areas.

Nonlinear Absorption The process wherein the attenuation coefficient becomes a function of the light intensity.

Photoablation The complex process phenomenon which includes intense UV light absorption, electronic heating effects, and explosive expansion, including shock wave movement, from a surface with only small residual heat deposited in the bulk of the sample.

Reading Bibliography

1. Tanaka, K., Miyajima, T., Shirai, N., Quan, Z., Nakata, R., "Laser Photochemical Ablation of $CdWO_4$ Studied with the Time-of-Flight Mass Spectrometric Technique," Journal of Applied Physics, Vol. 77., No. 12, p. 6581-7, June 1995.

2. Wolff-Rottke, B., Ihlemann, J., Schmidt, H., Scholl, A., "Influence of the Laser-Spot Diameter on Photo-Ablation Rates," Applied Physics A (Materials Science Processing), Vol. A60, No. 1, p. 13-17, Jan 1995.

3. Hahn, D.W., Ediger, M.N., Pettit, G.H., "Ablation Plume Particle Dynamics During Excimer Laser Ablation of Polymide," Journal of Applied Physics, Vol. 77, No. 6, p. 2759-66, March 1995.

4. Baeuerle, D., Arenholz, E., Heitz, J., Proyer, S., Stangl, E., Lukyanchuk, B., "Surface Patterning and Thin-Film Formation by Pulsed-Laser Ablation," Materials Science Forum, Vol. 173-174, p. 41-52, 1995.

5. Key, P.H., Sands, D., Wagner, F.X., "UV Excimer Laser Ablation Patterning of II-VI Compound Semiconductors," Materials Science Forum, Vol. 173-174, p. 59-66, 1995.

6. Fuso, F., Arimondo, E., "Ionization and Heating Processes During Laser Ablation and Vaporization of Solid Targets," AIP Conference Proceedings, No. 329, p. 97-100, 1995.

7. Hozumi, M., Murata, Y., Cheng-Huang, L., Imasaka, T., "Supersonic Jet/Multiphoton Iionization Spectrometry of Chemical Species Resulting from Thermal Decomposition and Laser Ablation of Polymers," AIP Conference Proceedings, No. 329, p. 81-6, 1995.

8. Zheng, R., Campbell, M., Ledingham, K.W.D., Jia, W., Scott, C.T.J., Singhal, R.P., "The Characterization of Laser Ablation of Bulk Super-Conducting Material Using Post Ablation Ionization of the Neutrals," AIP Conference Proceedings, No. 329, p. 353-6, 1995.

2.5. PRACTICAL CHARACTERISTICS OF UV REACTIONS

9. Gonzalo, J., Vega, F., Afonso, C.N., "Plasma Expansion Dynamics in Reactive and Inert Atmospheres During Laser Ablation of $Bi_2Sr_2Ca_1Cu_2O_{7-y}$," Journals of Applied Physics, Vol 77, No. 12, p. 6588-93, June 1995.

10. Toth, Z., Hopp, B. Kanotr, Z. , Ignacz, F. Szoerenyi, T., Bor, Z., "Dynamics of Excimer Laser Ablation of Thin Tungsten Films Monitored by Ultrafast Photography," Applied Physics A (Materials Science Processing)., Vol A60, No. 5, p. 431-6, May 1995.

11. Witanachchi, S., Mukherjee, P., "Role of Temporal Delay in Dual-Laser Ablated Plumes," Journal of Vacuum Science and Technology A (Vacuum, Surface, and Films), Vol. 13, No. 3, Part 1, p. 1171-4, May-June 1995.

12. Luby, S., Majkova, E., Illekova, El, Sandrik, R., D'Anna, E., Luches, A., Perrone, A., Enzo, S., "Effect of Laser Repetition Rate on the Melting and Ablation of Ni_24Zr_76 Alloy Ribbon," Thin Solid Films, Vol. 261, No. 1-2, P. 154-9, June 1995.

13. Blanchet, G.B., Fincher, C.R., Jr., "Laser Ablation as a New Tool for Material Science," Proceedings of the SPIE - The International Society for Optical Engineering, Vol. 1835, p. 2-12, 1993.

14. Vega, F., Afonso, C.N., Solis, J., "Real Time Optical Diagnostics of the Plume Dynamics During Laser Ablation of Germanium in an Oxygen Environment," Journal of Applied Physics, Vol. 73, No. 5, p. 2472-6, March 1993.

15. Anisimov, V.N., Der5kach, O.N., Kolomyiskii, Y.R., Nevmerszhitskii, V.I., Novobrantsev, I.V., Sebrant, A.Y., Stepanova, M.A., Yakovlev, E.R., "The Structure of the Plasma Plume, Produced in a Gas Under Atmospheric Pressure at the Surface of a Solid Target Irradiated by the XeCl Laser," Laser Physics, Vol. 1, No. 2, p. 188-95, March-April 1991.

16. Herman, P.R., Chen, B., Moore, D.J., Canaga-Retnam, M., "Photoablation Studies of Polymers, Quartz and Semiconductors with Vacuum Ultraviolet Laser Radiation," Photons and Low Energy Particles in Surface Processing Symposium, p. 53-8, 1992.

17. Pettit, G.H., Suaerbrey, R., "Pulsed Ultraviolet Laser Ablation," Applied Physics A (Solids and Surfaces), Vol. A56, No. 1, p. 51-63, Jan. 1993.

18. Demkovich, P.A., Kotlyar, M.L., Kulagin, V.Y., Sonin, K.V., "Modification of the Optical Properties and Dynamics of Laser Ultraviolet Ablation of Polyimide," Soviet Journal of Quantum Electronics, Vol. 22, No. 5, p. 400-4, May 1992.

19. Nakata, T., Kannari, F., Obara, M., "Ultraviolet and vacuum-ultraviolet laser ablation of organic polymers," Optoelectronics - Devices and Technologies, Vol. 8, No. 2, P. 179-90, 1993.

20. Srinivasan, R., "Ultraviolet laser processing of polymeric materials," Proceedings of the Symposia on Reliability of Semiconductor Devices, P. viii+330, P. 225-32, 1993.

21. Srinivasan, R., "Ablation of polyimide (Kapton/sup TM/) films by pulsed (ns) ultraviolet and infrared lasers," Applied Physics A (Solids and Surfaces), Vol. A56, No. 5, P. 417-23, 1993.

22. Srinivasan, R., "Ablation of polymethyl methacrylate films by pulsed (ns) ultraviolet and infrared lasers: A comparative study by ultrafast imaging," Journal of Applied Physics, Vol. 73, No. 6, P. 2743-50, 1993.

23. Babin, A.A., Bityurin, N.M., Polyakov, A.V., Fel'dshtein, F.I., Khvotova, N.L., "Nd laser fifth harmonic action on polymer films," Laser Physics, Vol. 2, No. 5, P. 805-10, 1993.

24. Herman, P.R., Chen, B., Moore, D.J., Canaga-Retnam, M., "Photoablation studies of polymers, quartz and semiconductors with vacuum ultraviolet laser radiation," Photons and Low energy Particles in Surface Processing Symposium, P. xv+552, P. 53-8, 1993.

25. Wada, S., Tashiro, H., Toyoda, K., Niino, H., Yabe A., "Direct photoetching of polymer films using vacuum ultraviolet radiation generated by high-order anti-Stokes Raman scattering," Applied Physics Letters, Vol. 62, No. 3, P. 211-13, 1993.

26. Pettit, G.H., Sauerbrey, R., "Pulsed ultraviolet laser ablation," Applied Physics A (Solids and Surfaces), Vol. A56, No. 1, P. 51-63, 1993.

2.5. PRACTICAL CHARACTERISTICS OF UV REACTIONS

27. Cemkovich, P.A., Kotlyar, M.L., Kulagin, V.Uy, Somin, K.V., "Modification of the optical properties and dynamics of laser ultraviolet ablation of polyimide," Kvantovaya Elektronika, Moskva, Vol. 19, No. 5, P. 441-6, 1992.

28. Srinivasan, R., "Etching polymer films with ultraviolet laser pulses of long (10-400 μsec) duration," Journal of Applied Physics, Vol. 72, No. 4, P. 1651-3, 1992.

29. Cain, S.R., Burns, F.C., Otis, C.E., "On single-photon ultraviolet ablation of polymeric materials," Journal of Applied Physics, Vol. 71, No. 9, P. 4107-17, 1992.

30. Haglund, R.F., Jr., Affatigato, M., Arps, J.H., Tang, K., Neihof, A., Heiland, W., "Ultraviolet laser ablation of halides and oxides," Nuclear Instruments & Methods in Physics Research, Section B, Vol. B65, No. 1-4, P. 206-11, 1992.

31. Leung, W.P., Tam, A.C., "Noncontact monitoring of laser ablation using a miniature piezoelectric probe to detect photoacoustic pulses in air," Applied Physics Letters, Vol. 60, No. 1, P. 23-5, 1992.

32. Stark, E.F., Laube, S.P., "Self-directed control of pulsed laser desposition," Journal of Materials Engineering and Performance, Vol. 2, No. 5, P. 721-26, 1993.

33. Dutta, S.K., Du, D., Squier, J., Pronko, P.P., Singh, R.K., "Laser ablation and materials response effects in semiconductors and metals for subnanosecond pulses," Proceedings of the Conference on Laser and Electro-Optics, Meeting v8, 1994.

34. Mukherjee, P., Sakthievel, P., Ahmed, K., Witanachchi, S., "Universality of ionic temporal bifurcation in laser-ablated plumes, Conference on Laser and Electro-Optics, Meeting v8, 1994.

35. Ueno, Y., Fujii, T., Inoue, S., Kennari, F., "Deposition of fluoropolymer thin films by vacuum-ultraviolet laser ablation," Proceedings of the Conference on Laser and Electro-Optics, Meeting v 8, 1994.

36. Phillips, Harvey, Feurer, T., LeBlanc, S.P., Callahan, D.L., Sauerbrey, R., "Excimer laser induced permanent electrical conductivity and nanostructures in polymers," SPIE, Vol. 1856, 1993.

item Matienzo, L.J., Zimmerman, J.A., Egitto, F.D., "Surface modification of fluoropolymers with vacuum ultraviolet irradiation," Journal of Vacuum Science & Technology, Vol. 12, No. 5, P. 2662-71, 1994.

37. Reyna, L.G., Sobehart, J.R., "Laser ablation of multilayer polymer films," Journal of Applied Physics, Vol. 76, No. 7, P. 4367-712, 1994.

38. Skumanich, A., "Surface light-induced changes in thick polymer films," SPIE, Vol. 2042, P. 146-53, 1994.

39. Dyer, P.E., Karnakis, D.M., "High quality ArF laser ablation of photoincubated polyethylene," Applied Physics Letters, Vol. 64, No. 11, P. 1344-6, 1994.

40. Bozhevolnyi, S.I., Potemkin, I.V., "Dynamics of ablation of polymethyl methacrylate films by near ultraviolet light of an argon laser," Journal of Physics D. (Applied Physics), Vol. 29, No. 1, P. 19-24, 1994.

41. Lumpp, J.K., Coretsopoulos, C.N., Allen, S.D., "Fluence dependence of excimer laser ablation of AIN," Electronic Packaging Materials Science VII Symposium, P. xiii+450, P. 213–18, 1995.

42. Liu, J.M., Zhu, S.N., Liu, Z.G., Zhu, Y.Y., Wu, Z.C., Ming, N. B., "Excimer laser ablating preparation of Barium/Sodium/Niobium Optical waveguiding films on Potassium/Ti Oxide/Phosphate substrates," Solid State Communications, Vol. 93, No. 6, P. 479-82, 1995.

43. Latsch, S., Hiraoka, H., "Diamond films by the excimer laser photoablation of polymers," JOM, Vol. 46, No. 7, P. 64-5, 1995.

44. Tsunekawa, M., Niship, S., Sato, H., "Laser ablation of polymethylmethacrylate and polystyrene at 308 nm; demonstration of thermal and photothermal mechanisms by a time-of-flight mass spectroscopic study," Journal of Applied Physics, Vol. 76, No. 9, P. 5598-600, 1995.

45. Itro, K., Moriyasu, M., "Photoablating characteristics of polymers irradiated by excimer laser with high repetition rates," Microelectronic Engineering, Vol. 25, No. 2-4, P. 305-12, 1994.

46. Nishikawa, Y., Yoshida, Y., Uesugi, Y., Mizunguchim, S., "Laser ablation mechanism deduced from structural changes of metal target," Journal of the Japan Institute of Metals, Vol. 50, No. 8, P. 936-43, 1994.

2.5. PRACTICAL CHARACTERISTICS OF UV REACTIONS

47. Hahn, D.W., Pettit, G.H., Ediger, M.N., "Optical properties of polyimide during ArF excimer laser ablation," Journal of Applied Physics, Vol. 76, No. 3, P. 1830-2, 1994.

48. Lash, J.S., Gilgenbach, R.M., Ching, C.H., "Laser-ablation-assisted-plasma discharges of aluminum in a transverge-magnetic field," Applied Physics Letters, Vol. 65, No. 5, P. 531-3, 1994.

49. Preuss, S., Matthias, E., Stuke, M., "Sub-picosecond UV-laser ablation of Ni films: strong fluence reduction and thickness-independent removal," Applied Physics A (Solids and Surfaces), Vol. A59, No. 1, P. 79-82, 1994.

50. Radharkrishnan, G., "Excimer laser ablation of contaminated polyimide (packaging)," SPIE, Vol. 2045, P. 40-6, 1994.

51. Yabe, A., Niino, H., "Polymer ablation with excimer lasers," Molecular Crystals and Liquid Crystals, Vol. 224, P. 111-121, 1994.

52. Chiba, A., Watakabe, Y., "Dynamic thermal response of photomasks caused by excimer laser pulse," Japanese Journal of Applied Physics, Part 1, Vol. 33, No. 1A, P. 341-6, 1994.

53. Zumoto, N., Izumo, M., Yagi, T., Myoi, Y., Tanaka, M., "Energy efficient optical system for excimer laser ablation apparatus," SPIE, Vo. 1869, P. 26-33, 1994.

54. Otis, C.E., Braren, B., Thompson, M.O., Brunco, D., Goodwin, P.M., "Mechanisms of excimer laser ablation of strongly absorbing systems," SPIE, Vol. 1856, P. 132-42, 1994.

55. Srinivasan, R., "A comparative study of the ablation of polyimide (Kapton) films by pulsed (ns) ultraviolet and infrared (9.17 mu m) lasers," AIP Conference Proceedings, No. 288, P. 439-43, 1994.

56. Ventzek, P, Gilgenbach, R., Ching, Chi H., Lindley, R.A., "Spacies-resolved laser-probing investigations of the hydrodynamics of KrF excimer and copper vapor laser ablation processing of materials," SPIE, V. 1856, 1993.

57. Pettit, G.H., Ediger, M.N., Hehn, D.W., Brinson, B.E., Sauerbrey, R., "Transmission of polyimide during pulsed ultraviolet laser irradiation,

Applied Physics A: Solids and Surfaces, Vol. 58, No. 6, P. 573-579, 1994.

58. Kukreja, L.M., Hess, P., "Photoacoustic detection of the decomposition kinetics of polymers: interpretation of acoustic signals," Applied Surface Science, Vol. 79, Pt 80, No. 1-4, P. 399-402, 1994.

59. Kukreja, L.M., Hell, P., "Time evolution of laser-induced polymer ablation studied by attenuation of a probe HeNe laser beam," Applied Surface Science, Vol. 79, Pt. 80, No. 1-4, P. 158-64, 1994.

60. Kumuduni, W.K., Nakata, Y., Okada, T., Maeda, M., "Spatial distribution of YO molecules ejected from laser-ablated film," Spplied Physics B, Photophysics and Laser Chemistry, Vol. 58, No. 4, P. 289-94, 1994.

61. Guzzo, E., Preston, J., "Laser ablation as a processing technique for metallic and polymer layered structures," IEEE Transactions on Semiconductor Manufacturing Vol. 7, No. 1, P. 73-8, 1994.

62. Germain, C., Girault, C., Gisbert, R., Aubreton, J., Catherinot, A., "KrF laser photo-ablation of a graphite target: Application to the development of thin films," Diamond and Related Materials, Vol. 3, No. 4-6, P. 598-601, 1994.

63. Tokarev, V., Marine, W., Lumney, J.G., Santis, M., "Time-resolved measurements of plume shielding during Arf laser ablation of silicon," Thin Solid Filoms, Vol. 241, No. 1-2, P. 129-33, P. 1994.

64. Marine, W., Tokarev, V., Gerri, M., Sentis, M., Fogarassy, E., "Ablation dynamics of silicon based targets inoxygen and nitrogen atmospheres," Thin Solid Films Vol. 241, No. 1-2, P. 103-108, 1994.

65. Srinivasan, R., "Pulsed ultraviolet laser interactions with organic polymers-dependence of mechanism upon laser power," Polymer Degradation and Stability, Vol. 43, No. 1, P. 101-107, 1994.

66. Ooie, T., Miyamoto, I., Maruo, H., Hirota, Y., Matsuyoshi, T., Orii, Y., "Ablation characteristics of material by excimer laser under different gas pressure and species," SPIE, Vol 1990, 1993.

2.5. PRACTICAL CHARACTERISTICS OF UV REACTIONS

67. Bor, Z., Hopp, B., Marton, Z., Gogolak, Z., Ignacz, F., "Ultrafest photography of shock waves originating from UV photoablated surface," SPIE, Vol 1983, P. 748-450, 1993.

68. Otis, C.E., Braren, B., Thompson, M.O., Brunco, D., Goodwin, P.M., "Mechanisms of excimer laser ablation of strongly absorbing systems," SPIE, Vol 1856, P. 132-42, 1994.

69. Nakamiya, T., Ikegami, T., Ebihara, K., "Dynamics during pulsed laser ablation of high T_c superconductor." IEEE Transactions on Applied Supercontivity, Vol. 3, No. 1, Pt., 3, P. 1057-60, 1993.

70. Xuejin, H., Huaimin, G., Yongrong, C., Mei, K., Tinghai, Y., Juxiong, X., Zhensheng, Z., "Research on laser ablation of various materials," Chinese Journal of Lasers, Vol. 19, No. 11, P. 876-80, 1993.

71. Pettit, G.H. Sauerbrey, R., "Pulsed ultraviolet laser ablation," Applied Physics A (Solids and Surfaces), Vol. A56, No. 1, P. 51-63, 1993.

72. Kuper, S., Brannon, J., Brannon, K., "Threshold behavior in polyimide photoablation: single-shot rate measurements and surface-temperature modeling," Applied Physics A (Solids and Surfaces), Vol. A56, No. 1, P. 43-50, 1993.

Chapter 3

The Excimer Laser

3.1 Introduction

Excimer lasers are a family of high pressure, pulsed gas lasers which produce intense UV light with high efficiency and high peak power at several useful wavelengths. The source of the emission is a fast electrical discharge in a high pressure mixture of a rare gas (argon, krypton, zenon) and a halogen gas (fluorine or hydrogen chloride). The specific combination of the rare gas and the halogen determines the output wavelength of the laser. There are four common wavelengths: 193, 248, 308, and 351 nm.

The excimer laser is the best known practical source of intense deep-UV radiation. The short wavelength UV energy from excimer lasers causes high levels of electronic excitation in irradiated materials leading to bond breaking and photoablation. The short wavelengths of the ultraviolet spectrum also provide a means for very high resolution imaging. These and other unique properties of UV light have provided a very broad field of applications for excimer lasers which are discussed in other chapters of this text. In this chapter we will cover the history, theory of operation, basic design, installation, subsystems operation, maintenance, and safety aspects of the excimer laser.

3.2 History

Excimer lasers were conceived in the 1960s and first demonstrated in 1972. In 1975 rare gas halogen (RGH) systems based on krypton fluoride (KrF) and xenon chloride (XeCL) were shown and in 1976 the first commercial excimer lasers became available. A schematic of an excimer laser gain generator is shown in Figure 3.1, illustrating the basic concept of operation. Depending on the types of lasers, the gain generator can be a diode (semiconductor laser), a solid rod of crystal such as a ruby or ND:YAG, a liquid as in a dye laser, or, in the case of the excimer, a gas. Other gas lasers include nitrogen, carbon dioxide, helium-neon, and argon.

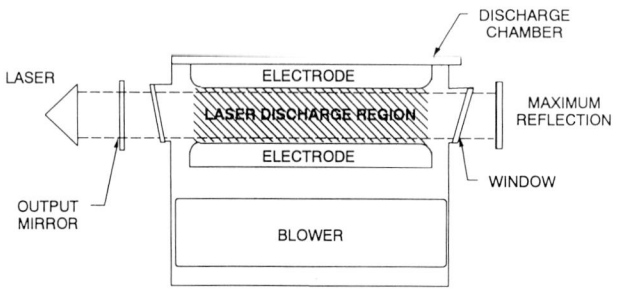

Figure 3.1: Schematic of an Excimer Laser Gain Generator

Excimer is an acronym for EXCIted diMER, a rare gas dimer based on one of the rare gas halogens, which are argon fluoride (ArF), krypton chloride (KrCL), krypton fluoride (KrF), xenon chloride (XeCL), xenon fluoride (XeFL), and xenon bromide (XeBr). Excimers (or exciplexes) are molecules with a very short lifetime, existing either in the excited state or in a weakly bound, dissociative ground state. Excimer molecules are produced by electron excitation in a gas mixture made up of fluorine or chlorine, a buffer gas (generally helium), and a rare gas (argon, krypton, xenon). The excimers, their characteristic lasing wavelength, and energy per photon are shown in Table 3.1.

3.3 Theory of Operation

The excited diatomic rare gas halide molecules mentioned above are responsible for the generation of UV photons which, increasing in the gain generator, produce the lasing action and a beam which exits from the output coupler. These molecules are either unbound or weakly bound in their electronic ground state. The actual excimer is formed by a chemical reaction

3.3. THEORY OF OPERATION

	λ (nm)	E (ev)
ArF	193	6.42
KrCl	222	5.50
KrF	248	5.00
XeCl	308	4.03
XeF	351	3.53

Table 3.1: Excimer Lasing Wavelength and Energy Per Photon

inside a pressurized laser discharge chamber where a UV preionized pulsed high voltage discharge produces the excited species.

The reactions begin with electron attachment to the fluorine molecule, followed by two-step ionization of the rare gas, then three-body excimer formation and spontaneous and stimulated emission. The spontaneous lifetime of an excimer is only a few nanoseconds, and parallel absorption reactions (photodissociation and photodetachment) are also occurring. The primary reaction channels are shown in Figure 3.2. The potential diagram which illustrates this reaction is shown in Chapter 2, page 44.

$F_2 + e \longrightarrow F^- + F$ Electron Attachment

$Kr + e \longrightarrow Kr^* + e$
$Kr^* + e \longrightarrow Kr^+ + 2e$ 2-step ionization

$Kr^+ + F^- + Ne \longrightarrow KrF^* + Ne$ 3-body Excimer Formation Reaction

$KrF^* \longrightarrow Kr + F + h\nu$ (spontaneous emission)
$KrF^* + h\nu \longrightarrow Kr + F + 2h\nu$ (stimulated emission) – LASER ACTION

$F + F + Ne \longrightarrow F_2 + Ne$ 3-body recombination (very slow)

Absorption Reactions

$F_2 + h\nu \longrightarrow 2F$ (Photo dissociation)
$F^- + h\nu \longrightarrow F + e$ (Photo detachment)

Figure 3.2: Primary Reaction Channels for Excimers

An excimer laser provides high power density discharge at high pressure capable of producing sufficiently high populations (10^{15} cm^{-3}) of excited species for suitable lasing action. The use of a transverse electrical discharge permits high reprates and efficient high average power. The uniform pumping of a relatively large volume of gas at high pressure is required to make a practical, commercial excimer laser.

The materials of construction that come in contact with the gas mixture need to be extremely inert so as to not react with the gas mixture and cause a loss of power. Molecular fluorine is highly reactive, highly corrosive, and toxic. By-products of corrosion in the laser discharge chamber will absorb UV light, deposit films and particles on the laser cavity optics, and interfere with the kinetics of laser excitation. Laser gas purity must also be very high for the same reasons. Impurities in the system will consume the halogen component of the gas, reducing gas lifetime.

3.4 The Gain Profile

The gain profile of a typical excimer laser is shown in Figure 3.3, taken from the side light fluorescence spectra of a discharge-pumped KrF laser. These data were taken on a Jobin Yvon THR 1500, 3-m spectrometer (1).

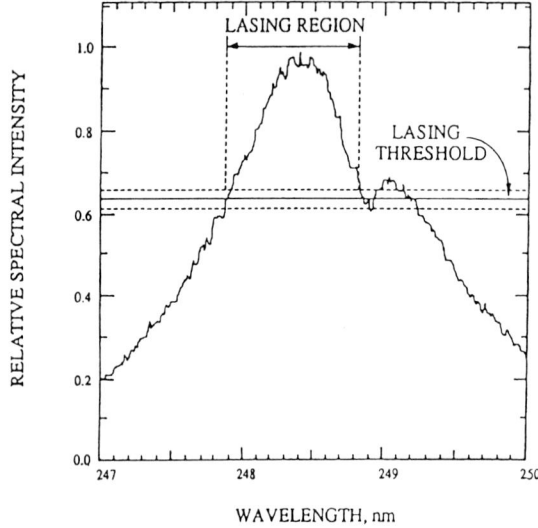

Figure 3.3: Gain Profile of a Krypton Fluoride Excimer Laser

The relative spectral intensity is indicated as a function of wavelength; the lasing region is between 247.9 and 348.7 nm. The main properties of the

3.5. EXCIMER LASER SUBSYSTEMS

laser that produce the spectral properties are the high pressure buffer gas, since pressure increases the number of three-body collisions; rapid pumping of highly energetic electrons (if pumping is reduced, arcs will form in the discharge due to lack of absorption of electrical energy by the gas); the low fluorine concentration (0.1%) is maintained since too much fluorine will consume needed electrons; high gain (0.15-0.2 cm^{-1}) is achieved, so much so that light is produced outside the laser even without mirrors; and there is no optical cavity buildup and low speckle as a result of the multimode (harmonic mixing) behavior of the light. Other characteristics of excimer lasers are pulsed operation and very short (7 nsec) upper state lifetimes.

3.4.1 Coherence Properties

Most lasers are extremely coherent, and coherence is an advantage in forming vertical structures without lateral effects. In microlithography, high coherence causes interference effects, such as speckle and standing waves, that degrade IC image quality and resolution. The excimer laser is characterized by good coherence without speckle, caused mainly by a large number of transverse modes in the beam output. The number of modes and the extent of wave mixing result in multimode beam characteristics. Sufficient incoherence eliminates the problem of speckle.

3.5 Excimer Laser Subsystems

An excimer laser is a system composed of distinct subsystems. The primary subsystems are 1) the laser head module; 2) the electrical energy transfer system; 3) the optical resonator; 4) the control system; and 5) support systems. We will discuss each of these in detail.

3.5.1 Laser Head Module

The laser head module is basically the discharge chamber, which includes the laser cavity, electrodes, preionizer system, blower, heat exchanger, windows, and a filtration system to remove metal fluoride particles caused by the ablation of electrode material.

Various methods of excitation are used in excimer lasers, but most are preionized with a UV corona followed by an avalanche electric discharge. Other methods of excitation are direct electron beam (which is characterized by high pulse energy, low reprates, and low component lifetime) and microwave/rf excitation which has low pulse energy with high repetition rates and low average power.

UV preionization schemes provide a direct source of UV photons to help initiate a uniform, constant discharge and reduce peak voltage. A corona preionizer (shown in Figure 3.4) generates a low level of UV photons which in this example pass through a perforated cathode into the discharge path.

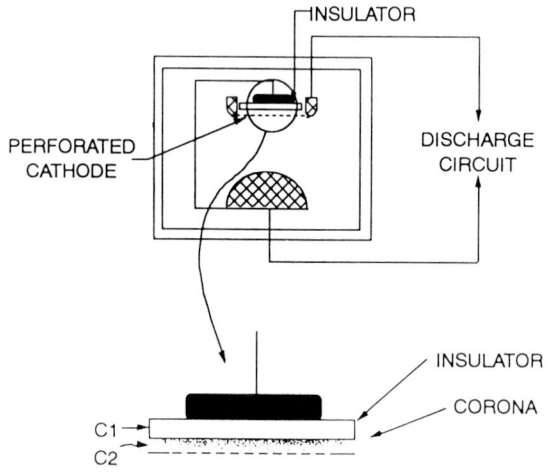

Figure 3.4: Schematic of a Corona Preionizer

The corona preionization is moderately (1%) efficient and supports reprates in the 100-to 500-Hz range and energy levels of 100 mJ - 1 J. Corona preionizers are relatively inexpensive and easy to build to production quantities.

The avalanche electric discharge of an excimer laser has a relatively fast discharge circuit (T 65w) at high voltage (15-25 kV) and high current (30-50 kA) discharge. It is also typically a low impedance discharge (0.50 hm) with high energy loading (100-200 J/l) and pulse durations in the 10- to 30-nsec range. The actual discharge chamber or laser head is depicted in Figure 3.5.

A high speed blower moves the gas volume rapidly between the electrodes while excess heat is removed by either water or air cooling. The metal fluoride trap which removes the particular by-products of gas corrosion and ablated electrode material is not shown here.

3.5. EXCIMER LASER SUBSYSTEMS

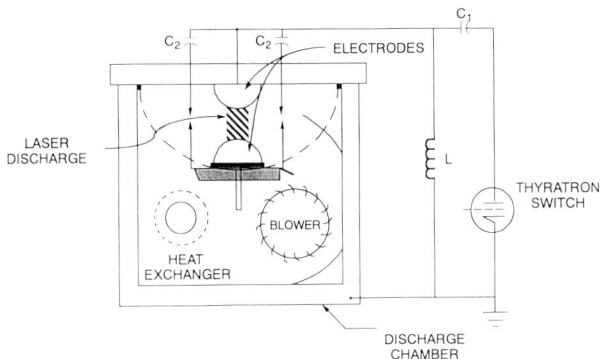

Figure 3.5: Excimer Laser Discharge Chamber Schematic

3.5.2 Optical Resonator System

The optical resonator subsystem manages the direction of the light out of the laser cavity and includes a maximum reflector positioned at the rear of the laser (RMAX) and a partial reflector at the front of the laser which both reflect light back to RMAX and at a given intensity couple light out through the laser output coupler. A schematic of an optical resonator is shown in Chapter 1.

The laser may also incorporate a wave meter to monitor output wavelength, as well as a support structure which contains the optics and physically isolates this subsystem from the influences of vibration and heat created by the discharge chamber.

3.5.3 Control System

The control system manages the distribution of power throughout the laser (power supplies, discharge circuit, blower motor, diagnostic sensors, valves, etc.) with a computer and a controller or I/O module custom configured to the various laser functions. Multiple lasers can be networked or tied to individual optical delivery systems which carry their own host computer. Lasers may also be remoted behind chase walls containing plumbing and a gas supply, and related facilities support them in order to isolate the system from a clean room or an atmosphere-controlled process area. The laser is controlled from a separate module which is located in a clean room or another temperature or particle controlled environment. This type of "remote" operation is typically used in wafer stepper application.

3.5.4 Support Systems

Support modules for excimer lasers include a variety of devices to ensure safe, uninterrupted production operation of the system. The gas handling system delivers excimer laser gas mixtures from source cylinders through a system of valves to the laser input manifold. Gas bottles are either contained in vented cabinets or integrated in a vented enclosure with the laser and its optical delivery system.

Spent gas is pumped out through a fluorine trap, which may contain a sight glass to indicate saturation of the trap. Excimer laser gas in some lasers is circulated through a cryogenic procession to freeze out particles of corrosion that, if left in the system, reduce gas life. Other methods for removing metal fluoride particles include electrostatic precipitators and similar dust traps. Figure 3.6 shows a typical gas plumbing facility schematic (courtesy of Cymer Laser Technologies).

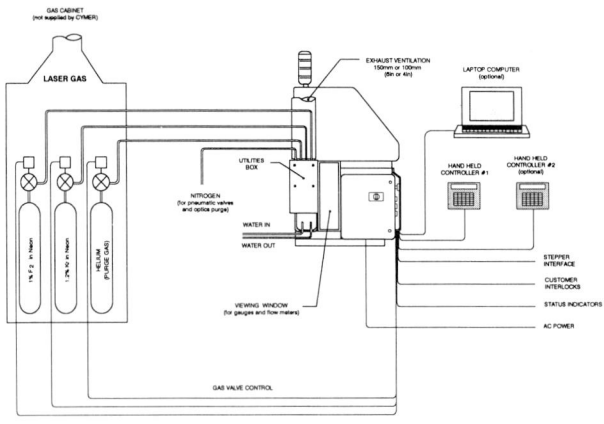

Figure 3.6: Excimer Laser Gas Facility Schematic

Excimer laser systems require electrical or electromechanical interlocks so that human access is prevented while the laser is on or shutters are activated to block the beam during limited operator access to the system. All high voltage components are similarly protected by safety interlocks, and the overall system (laser and beam delivery system) is shut down with a single, large mushroom-type button called an "EMO" or emergency off switch.

Water cooling systems, blower motor and power supplies, and various venting systems and fans to circulate fresh air around the inside of an excimer laser enclosure will remove gas in the event of a leak. Gas and air monitors, laser sensors, and room door interlocks ("laser on" signs) are examples of support systems.

3.6 Laser Operation

An initial step in operating an excimer laser, whether for research or production applications, is checking its performance against the manufacturers published specifications. Generally, excimer lasers are given a short "burn in" at the manufacturer's site where they are operated for several million shots. At this stage, the basic operating parameters are logged and documented as "acceptance test data." The data sets from these tests are shipped with the laser. Following installation, a second similar set of tests are run to verify the operating performance, and at regular intervals the laser is tested again to measure its performance over many millions of shots.

In this section, we will review the initial performance specification tests and present typical preventative maintenance testing and recording. Maintenance procedures with specified intervals are used to keep the power levels up, and component lifetimes are also predicted by the number of laser pulses so that they can be replaced before failing.

3.6.1 Spectral Output

The optical output of the laser is one of the first properties to be measured. In many applications, the "raw" or natural, broadband output is used without any spectral narrowing since optical color or bandwidth is not an issue in micromachining and applications. In high resolution lithography, spectral narrowing to below 1 pm is required to keep image focus and other optical properties within specifications.

76 CHAPTER 3. THE EXCIMER LASER

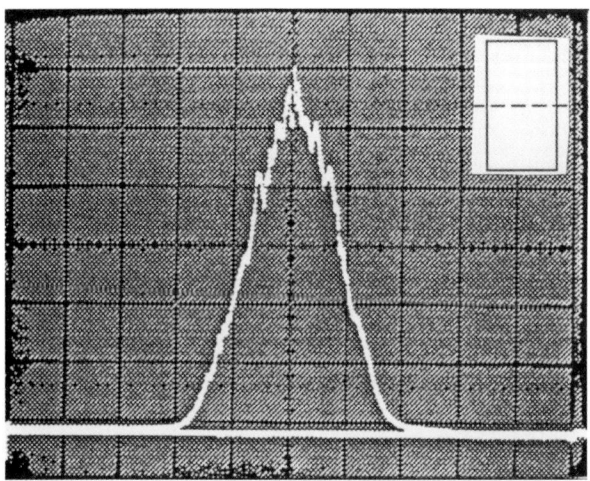

Figure 3.7: Horizontal Beam Scan: Excimer Laser at 248 nm

3.6.2 Beam Intensity Profiles

Many different tests are used to measure the performance of an excimer laser, one of the primary being beam intensity profiles, which are a direct measure of the total energy coming from the laser. Figures 3.7 and 3.8 show horizontal and vertical beam scans taken with a linear photodiode array. The top-hat shape on the left is from a vertical slice of the beam, and the Gaussian shape on the right is from a horizontal slice. The scale is 1 div = 2.5 mm of the beam (5); laser 400 Hz, 14 kV, 1 m from the output.

The profiles of laser beam intensity express the cross-sectional (vertical and horizontal) footprint area of UV light being transmitted to the optical delivery system. These intensity profiles are critical to understanding how to optimally utilize all of the available UV energy when transferring the laser beam into an optical delivery system. Apertures are used to clip the nonuniform areas of the beam, leaving the central "sweet spot," the area of greatest energy, to optically shape and transmit through the delivery.

Beam scans are taken at some fixed distance from the output end of the laser (1) at a point centered on the photodiode array. The laser beam intensity profiles are used to diagnose the internal "health" of the discharge chamber. For example, if the electrodes wear unevenly, the beam profiles

3.6. LASER OPERATION

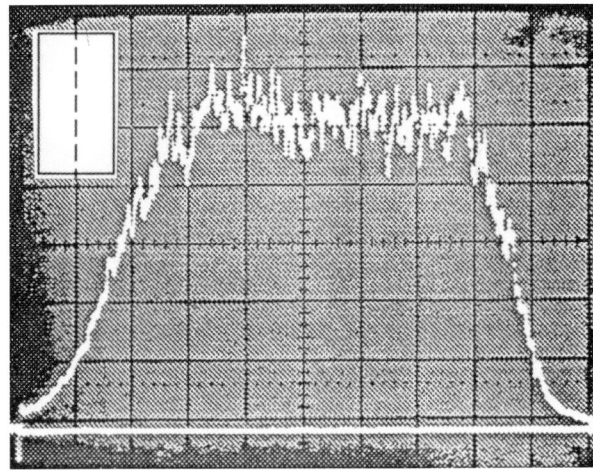

Figure 3.8: Vertical Beam Scan: Excimer Laser at 248 nm

will change and indicate the problem.

3.6.3 Pulse Characteristics

Pulse durations in excimer lasers range from picoseconds to several microseconds, depending on the pumping source and component configuration. Longer discharge excitation pulses are used to improve beam quality. X-ray preionization is also used for longer pulses. Pulse compression, on the other hand, is desirable in applications of laser fusion and avalanche discharge preionization may be used in this application. The scaling of the excimer laser to much longer pulses and higher output power is somewhat restricted by arching in the discharge chamber. However, the real need for increased output power results in large, high power (1- to 5-kW) excimer lasers.

The most commonly used excimer lasers operate with pulse lengths of 15-45 nsec, using UV preionization to initiate the discharge. A key advantage of the excimer laser over other laser sources is the short coherence length and broad spectral content that greatly reduces or eliminates interference effects such as speckle. The multimode behavior of excimer lasers results in a well-homogenized beam output. This is an important characteristic in the use of excimer lasers for IC processing operations. For example, film thickness control is measured in the 20- to 100-Å region. Speckle could cause major variations in resist imaging dimensions, and would eliminate the use of excimer lasers for photolithography. Speckle would also limit the use of excimer lasers in high resolution ablation and direct imaging applications. Figure 3.9 shows a typical excimer laser three-dimensional pulse profile, taken from a Lambda Physik xenon chloride laser with 50 pulses averaged.

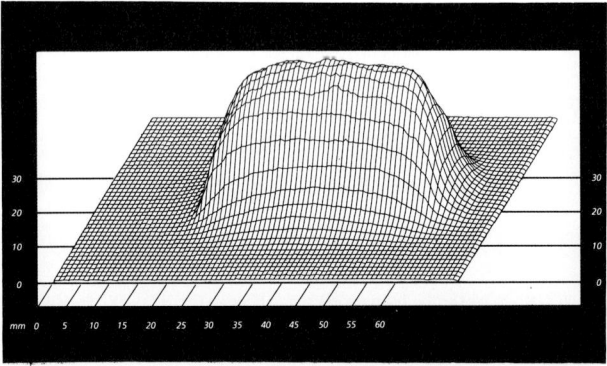

Figure 3.9: A Typical Excimer Laser Pulse Profile

3.6.4 Pulse-to-Pulse Stability

The individual pulses from an excimer laser vary according to the exact position of the electrical discharge and the internal gas and pressure conditions. Preionization also affects pulse characteristics. A plot of normalized pulse energy and number of pulses taken from a Cymer CX Series laser is shown in Figure 3.10.

Over a 5000-shot burst, the one sigma variation is only 1.42% at 200 Hz. Data were taken on an EG&G Gamma Scientific radiometer and analyzed on a lab computer. Stabilized pulse energy is an important specification for an excimer laser. Pulse energy multiplied by the reprate (Hz) determines the average power, generally the most important specification in determining the suitability of a given laser for a specific application.

3.6. LASER OPERATION

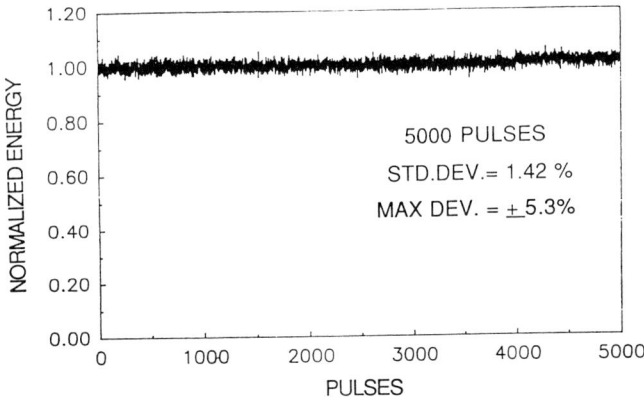

Figure 3.10: Pulse-to-Pulse Energy Variations

3.6.5 Gas Lifetime

Gas lifetime is determined by several factors, including the materials of construction for the discharge chamber, the operating parameters of the laser, and the consumption of fluorine. Static gas lifetime is the length of time the laser can operate at a fixed power level without refilling the chamber. Static gas lifetime can be with or without periodic injections of flourine to boost the output and keep the voltage below a specified maximum. As the fluorine is consumed, the laser operating voltage is increased to maintain a constant output power. At some point, the voltage becomes excessive (causes strain on the discharge circuit components), and the discharge chamber must be pumped out and fresh gas bled in. Figure 3.11 shows a typical gas lifetime chart.

3.6.6 Output Power and Reprate

The output power is critical in making the excimer laser an economical and productive tool for manufacturing applications. In many application, it is desirable to have a high repetition rate and low pulse energy. This will result in short "exposure" or dose intervals and good dose control (1-3%). High throughput and uniform exposure doses of energy are key manufacturing requirements for semiconductor application and are desirable for most production manufacturing processes.

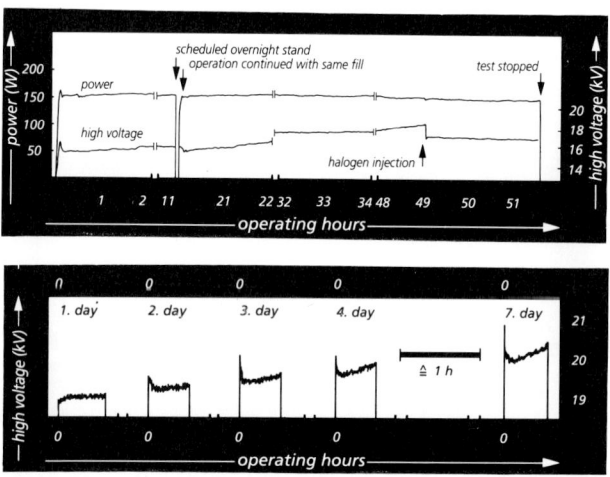

Figure 3.11: Static Gas Lifetime (7)

Typically, reprates for excimer lasers are in the 100- to 1000-Hz range and can be operated over this range by selection in the laser control system computer. A microprocessor is used to control individual pulses and the operating voltage as a means to adjust pulse energy. Energy locking and monitoring functions are performed by the computer to ensure uniform energy dosing from the laser to the delivery system.

3.6.7 Long Term Operation and Reliability

Long-term reproducibility of the excimer beam profile is important in most applications. UV-preionized excimer lasers are the "work horses" of the UV laser industry, and despite a high efficiency preionizer, electrodes are ablated with use, causing changes in pulse profiles. Electrode material and shape are very critical parameters for this reason, along with degree of preionization and discharge density. All of these must be carefully balanced with respect to each other.

Figure 3.12 shows six pulse profiles taken over various intervals as indicated. The abscissa is the electrode gap, the ordinate is the lateral dimension of the discharge, and the vertical represents intensity in arbitrary limits.

The laser used was operated at 100 Hz using a gas purifier and was rated

3.6. LASER OPERATION

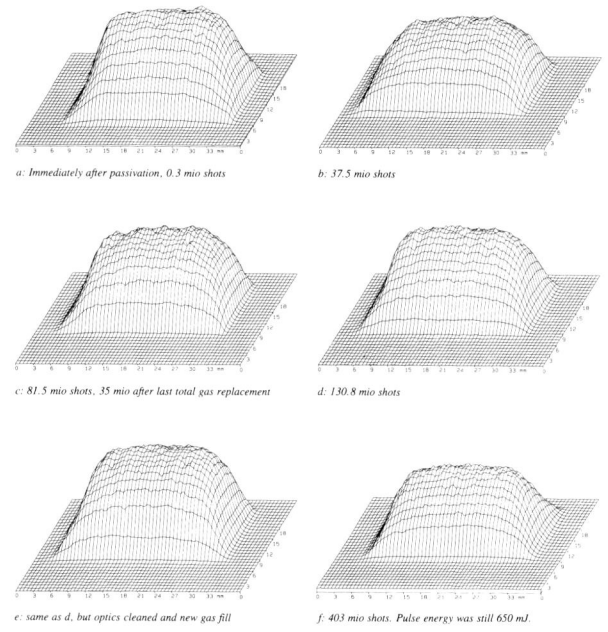

Figure 3.12: KrF Beam Profiles over 400 M Shot Test (8)

at 600 mJ. The power was stabilized by using halogen injections, voltage control, and a combination of partial and total gas replacement. One modification of the laser for this experiment was a reduced electrode gap. The important aspects to note are relative changes in the abscissa (electrode gap) and in relative intensity changes in various parts of the energy profiles. The high reliability studies are critical to making excimer lasers high volume manufacturing tools. Gas composition is the other variable (not part of the study) for high reliability operation.

3.7 Laser Installation

The installation of an excimer laser will require several gases, water for cooling, exhaust ventilation to purge the laser enclosure, and power. The gases required are usually either krypton (1% in neon) or argon (1% in neon) or xenon in neon. A typical premix recipe is 0.1% fluorine, 1.2% krypton, and the balance neon. Helium is used as a purge gas and filtered dry nitrogen is used for operating the pneumatic valves and purging the laser optics. A fully facilitized excimer laser includes the laser gas cabinet and gas lines, exhaust ventilation, nitrogen line for optics purging and pneumatic valve operation, AC power line, water lines, control system electronics, status indicators, interlocks, computer and controls, and an interface line to an optical delivery system.

3.8 Laser Maintenance

There are several key laser subsystems that require periodic maintenance when operating an excimer laser in a production environment. These include the laser gas, the laser windows, the metal fluoride trap, the laser chamber, and the pump-out system. Overall, the entire laser system also must be monitored, inspected, and calibrated periodically. Selected parameters can then be adjusted to avoid drift during use. In this section, we will discuss these various maintenance areas, their symptoms, causes, and cures.

3.8.1 Laser Gas

Laser output deteriorates as the gas is consumed in an excimer laser, caused mainly by the burning of fluorine due to continuous passivation of electrodes. Another cause of reduced UV output is the buildup of contamination caused

3.8. LASER MAINTENANCE

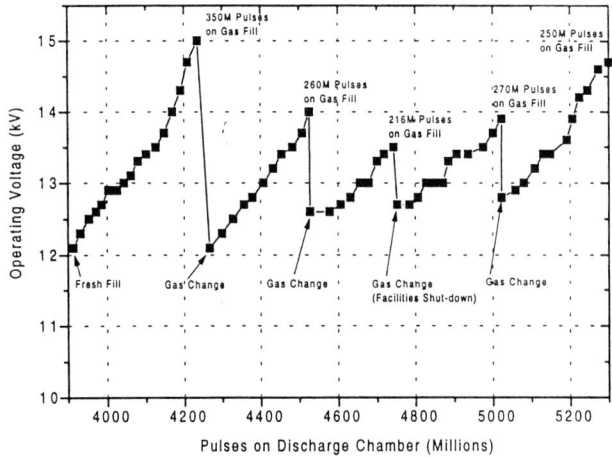

Figure 3.13: User Operating Voltage vs Pulses and Gas Change (8)

by fluorine reacting with materials in the discharge chamber. Fluorine reacts with these materials (plastics, metals) to form CF_4, SiF_4, HF, SF_6, and CF_2.

The typical drop-off in output power from gas-related causes is 5-25% in a 24-hr period or over several million pulses. Lasers are programmed for fluorine injection at regular intervals to maintain a minimum level of flourine atoms to sustain power output.

Gas lifetime is improved by the use of inert materials inside the discharge chamber to reduce the reactivity with fluorine. The high voltage discharge causes ablation of metal from the electrodes, and these particles also react with fluorine to reduce output power. A cryogenic gas processor can be used to freeze out these particles, and gas circulated through a "cryoprocesser" is used frequently for this purpose. Any techniques which allow the laser to run at lower voltage, taking stress off of the entire system (electrical system, discharge chamber parts), are advantageous, as shown in Figure 3.13.

3.8.2 Laser Windows

Laser windows can decrease laser output as particles in the circulating gas mixture deposit on the window surfaces and block UV photons. Dirty windows will also cause deterioration of the laser beam profile. Metal fluoride particles and the deposition of organic compounds outside the windows will degrade laser output. Also, the fluorine will gradually deteriorate the window by etching its surface.

There are several methods to improve the lifetime of laser windows. The most important is to filter out the particles that land on window surfaces by cryogenic processing of the gas, by electrostatic precipitation of the particles, or by some other means. Window surfaces are also purged with nitrogen gas to keep particulates from accumulating and are kept extremely dry since water mixes with fluorine to form hydrofluoric acid which will etch the windows.

Windows must be baked in an oven before installation in the laser. Window cleaning is performed by dragging a lens tissue across the surface of the window (gravity force only) with a sheet of lens tissue after wetting it with dichlorofluoroethane. This is solvent made by several companies, including Allied Signal (Genesolv 2000 is the product name).

The cleaning interval for windows varies according to the laser type and operating parameters. Typical window cleaning intervals range from as little as 200 million pulses to over 500 million pulses on excimer lasers which have the ability to operate at relatively low (<15-kV) voltages.

3.8.3 Laser Electrodes

Laser electrode wear will also cause a degradation of laser output (and bandwidth degradation for line-narrowed lasers). The electrodes wear unevenly in the discharge chamber, which also causes degradation in pulse-to-pulse energy stability and changes in the beam profile. Electrode wear occurs as the fluorine continually repassivates the electrode surfaces and as electrical arching ablates the metal surface.

The composition of the electrode is critical (many electrodes are solid brass or combinations of brass and other metals), and various alloys are used for low erosion characteristics. Electrode wear can be reduced by keeping the operating voltage low and by clever design of the electrode shape to permit more optimal and uniform surface erosion. Some electrodes are round, others are semiflat, and some are oval or elipse shaped.

3.8.4 Laser Power System

Excimer lasers, and other gas and solid state lasers, have power systems which require periodic preventative maintenance. The circuits of excimer lasers wear from the high voltage-induced stress and require some "spare

3.9. GAS SAFETY GUIDELINES

part" replacement from time to time. The thyratron switches in excimer lasers are a more common part requiring replacement, as they suffer from trigger board degradation, overheating, and just plain aging. A good thyratron should last for about 750-950 million pulses or more. Solid state thyratrons have a lifetime of several billion shots.

Misfiring is a common symptom for power system problems. An aging thyratron will cause misfiring. Ceramic and other electrical system components will also suffer from the stress and aging normal to high voltage systems. Components will crack and metal parts become fatigued over time and require replacement.

3.8.5 Laser Optics

Lasers that are line narrowed or contain other special optics can cause operation problems if not properly maintained. The most common problem with optics of this type is solarization or color center formation, leading to increased UV absorption and eventual cracking from stress and heat. It is most important to purge all the laser optics with a good flow of dry nitrogen. The optical materials used must be of high purity and well-coated to prevent UV laser beam damage. Impurities in the bulk of optical components act as nuclei for UV photons; these sites will become "attractors" for intense absorption, leading to heat, expansion, and ablation. The optical coatings face the same problems: absorption caused initially by imperfections (voids or inclusions/dirt) that cause absorption.

3.9 Gas Safety Guidelines

The main requirement for gas distribution is to have a safe, clean, and well plumbed system. This requires that standard techniques for safe gas handling are used, such as chaining all pressurized cylinders to walls and transporting property, use of the proper high quality fittings and strictly following CGA plumbing guidelines, and checking all gas lines for leaks periodically and, if all gas is stored in one room, providing an alarmed detector.

The actual gas plumbing must accommodate valves for complete purging and evacuation of lines, and all excimer gas lines must be completely cleaned of hydrocarbons and water or residual moisture. The tubing should be a low-carbon stainless steel which is electropolished.

The recommended ventilation flow rate for fluorine lines is approximately 175 feet over all plumbing connection joints and about 300 CFM venting of the laser enclosure. Fluorine is a class IV toxic gas with a TLV threshold limit valve or maximum allowable concentration (MAC) rating of 1 ppm.

The fluorine lines are typically a double-walled coaxial type for feed and exhaust. Spent gas is flushed through the fluorine trap up to a specified number of chamber volumes (usually 200-300). When the laser is taken "down" for maintenance, ensure that adequate air/ventilation is also provided for operator safety.

3.10 UV Safety

Ultraviolet light from an excimer laser is extremely intense radiation and exposure to a direct beam can cause severe burns to human tissue. The UV light is "Class IV Radiation." Please refer to the appropriate CDRH regulations. Also, refer to other relevant local state and municipal safety codes which may apply. In most countries, all laser beam delivery paths must be completely shielded, enclosed in a proper beam tube, and electrically or electromechanically interlocked.

The following information is taken directly from a safety sheet written at Spectra Gases Inc.:

Excimer Gas Hazards

> "A multitude of hazards are associated with handling laser gases, including high pressure, asphyxiation, toxicity, and corrosivity. The gases typically used vary with laser type and manufacturer. However, all excimer lasers have fill gases comprised of halogens including either HCl, F_2 or sometimes NF_3; rare gases including Xe, Kr or Ar; and a buffer gas of He and/or Ne.
>
> There are hazards common to all of the above gases but the one most overlooked and underestimated is the high pressure itself. All cylinders used in the laser industry have pressures typically between 300 psi and 3000 psi. At these pressures, an accident causing the cylinder to fall could shear an unprotected valve, causing the cylinder to simulate the action of an unguided missile. A second hazard also underestimated by laser gas users is found in the seemingly benign gases such as He, Ne and Ar which can act as suffocants due to the displacement of oxygen.
>
> The two hazards most often associated with the use of excimer

3.10. UV SAFETY

gases are the corrosivity and toxicity of the halogen component. Fluorine and hydrogen chloride gases are by far the most common halogen donors in excimers. Extreme care should be taken when handling these products. Both gases, if not properly used and safely stored, can be devastating. They are irritants of the upper respiratory tract and can cause pulmonary edema. Exposure to high concentrations of either gas, even if only brief, is usually fatal. Mild, dilute exposures will result in coughing and choking and burning of the eyes and nose. Vapor contact with the skin will cause tissue irritation and necrosis. Although not usually present in the pure form for this application, liquid HCl or F_2 can cause painful burns of the tissue and bone if they come in contact with the skin. These effects are real and are not overstated for effect.

Special first aid preparations for F_2 burns should be used immediately. These include magnesium oxide and glycerin paste, as well as a 10% calcium gluconate solution for injection under the skin (after use of local anesthetic) to halt tissue decay.

Another problem these halogens pose is the corrosive attack on laboratory equipment. Of the two gases, HCl is the more destructive. If an HCl mix were to vent accidentally, every piece of electronic equipment including computers and telecommunications equipment would be ruined. Optical components and other specialized equipment in a laser lab would also be adversely affected.

If precautions are not heeded, accidents can and will occur. If a rapid or violent release of any gas occurs from a cylinder not secured, contained and vented, one must quickly ascertain what went wrong and what laser gas is escaping into the room. Then make sure the ventilation system is on and that all personnel are removed from the area. Don't try to be a hero and try to stop the leak of a toxic gas.

If anyone was exposed to toxic gases or shows evidence of suffocation due to the lack of oxygen, administer first aid and call a physician immediately.

First aid for inhalation of fluorine, hydrogen chloride and even carbon dioxide laser mix, if it was severe enough to cause loss of consciousness, is nearly identical. If breathing is labored, give 100% oxygen under positive inhalation pressure for half-hour periods every hour for six hours. If the patient is not breathing, give artificial respiration. The patient should be kept warm, not hot, and stay under the supervision of a physician until the danger has passed.

First aid for halogen mixes which contact the skin and eyes is to subject the person immediately to a drenching shower with all clothing being removed as rapidly as possible. With fluorine exposure, the skin should be washed with 2 to 3 % aqueous ammonia solution and then flushed with water again. Compresses that include saturated solution of Epsom salts or iced 70% alcohol should be applied for at least 30 minutes as well. Check with your specialty gas supplier for more details. For HCl contact, flushing with water only is recommended. BE SURE TO RECEIVE THE ADVICE OF A QUALIFIED PHYSICIAN QUICKLY.

Make sure the gas has completely dissipated before re-entering the lab. After donning clean neoprene gloves, coat, boots and a self-contained breathing apparatus, enter the lab and inspect the damage. If it was a halogen gas that vented, remove the cylinder and then systematically clean away the films of hydrofluoric or hydrochloric acid that will have formed on all the equipment."

References

1. Sengupta, U.K., "Excimer Lasers," A Short Course Developed at Cymer Laser Technologies, San Diego, CA, 1992.

2. Pummer, H., "Commercial Excimer Lasers," Lambda Physik, Acton, MA, 1991.

3. C.K. Rhodes (Ed), "Excimer Lasers" Series: Topics in Applied Physics, Vol. 30, Springer Verlag, Berlin, 1979.

4. Sengupta, U.K., "Excimer Laser for Microlithography: Working Knowledge and Practical Considerations," SPIE Short Course, SC11, March, 1990.

3.10. UV SAFETY

5. Elliott, D.J., Pennelli, C.P., Sengupta, U.K., "Recent Advances in an Excimer Laser Source for Microlithography," J. Vac. Sci. Technol. B 9 (6), Nov/Dec 1991.

6. Lumonics Technical Publication "Excimer Lasers for Industry," 1993.

7. Internal Publication, Lambda Physik, 1993.

8. Lambda Physik Highlights Publication, 1993.

Glossary of Terms

Average Power In laser optical systems, the separate pulses per second (h_2) multiplied by the pulse energy.

Band A group of spectral lines in a molecule's absorption spectrum.

Chirp Time-varying frequency.

Coherence A condition of light occurring when the various phases and wavelengths of energy in a single beam are closely aligned.

Excimer Acronym for "excited dimmer," a molecule consisting of generally two atoms (for example, krypton and fluorine) which are strongly bound in the excited or upper energy level state and dissociate in the ground state; an excited complex of two molecules.

Exciplex Acronym for an excited gas complex, generally refers to two molecules.

Full Width, Half Maximum The points on an oscilloscope screen describing how laser pulse energy is measured, that being at the 50% point on the vertical scale (half maximum) and extending to the full left and right boundaries on the y-axis or "full width."

Free Spectral Range Frequency difference between adjacent resonances in an optical cavity.

Gain Medium The area inside a laser in which photons are amplified using mirrors at either end; selected molecular spaces between the mirrors which are excited to upper energy states.

Hertz A measurement unit for frequency; one hertz equals one cycle per second.

Inversion A laser with a gain medium containing ions in the gas discharge.

Jitter The average residual variation in the frequency of a single mode laser beam; referred to a frequency linewidth deviation.

Laser An acronym for light amplification by stimulated emission of radiation. A stimulated emission device that produces intense, highly coherent monochromatic optical radiation.

Laser Cooling The use of pressure from laser radiation to extract kinetic energy from atomic mution.

Laser Resonator The emission of ultraviolet (UV) or infrared (IR) radiation of excited species materials.

Linewidth Wavelength or frequency spread distance over which emission, absorption, and gain take place.

Mode A characteristic form of oscillation in a laser; can be transverse (across the medium) or longitudinal (along the medium).

Multimode Laser beams with two or more transverse modes.

Q Switching An optical shutter device in a laser cavity used to generate a short, high energy pulse by controlling the quality factor or resonant Q.

Resonator A cavity providing space for the lasing medium with mirrors at opposite ends.

Threshold The point where a laser has sufficient gain to permit oscillation.

Tunable The ability to adjust laser wavelength within a gain bandwidth.

Reading Bibliography

1. Takahashi, M., Maeda, K., Kitamura, T., Takasaki, M., Horiguchi, S., "Experimental Study of Formation Kinetics in a Discharge-Pumped F_2 Laser," Optics Communications, Vol. 116, No. 1-3, p. 269-78, April 1995.

3.10. UV SAFETY

2. Nagai, S., Furuhashi, H., Uchida, Y., Yamada, J., "FormationDynamics of Exicted Atoms in an ArF Laser Using He and Ne Buffer Gases," Journal of Applied Physics, Vol. 77, No. 7, p. 2906-11, April 1995.

3. Ishii, A., Yasuoka, K., Okita, Y., Tamagawa, T., "Three kHz XeCl Excimer Laser Using New Type of Electrode," Japanese Journal of Applied Physics, Part 1 (Regular Papers and Short Notes), Vol. 34, No. 5A, p. 2324-8, May 1995.

4. Borisov, V.M., Dmitriev, A.A., Prokofiev, A.V., Khristoforov, O.B., "Conditions for the Excitation of the Wide-Aperture XeCl Laser with an Average Output Radiation Power of 1 kW," Quantum Electronics, Vol. 25, No. 5, p. 408-10, May 1995.

5. Losev, V.F., Panchenko, Y.N., "Formation of High-Quality XeCl Laser Radiation in a Cavity with an SBS Mirror," Quantum Electronics, Vol. 25, No. 5, p. 450-1, May 1995.

6. Losev, V.F., Panchenko, Y.N., "Characteristics of Stimulated Scattering of Broad-Band XeCl Laser Radiation," Quantum Electronics, Vol. 25, No. 5, p. 448-9, May 1995.

7. Borisov, V.M., Borisov, A.V., Bragin, I.E. Vinokhodov, A.Y., "Effects Limiting the Average Power of Compact Pulse-Periodic KrF Lasers," Quantum Electronics, Vol. 25, No. 5, p. 421-5, May 1995.

8. Efimovskii, S.V., Zhigalkin, A.K., Kurbasov, S.V., "Gain Spectra of a Long-Pulse XeCl Laser, Measured in the Range 307.5-308.8 nm with a Resolution of 1 cm^{-1}," Quantum Electronics, Vol. 25, No. 5, p. 430-5, May 1995.

9. Makarov, M., "Effect of Electrode Processes on the Spatial Uniformity of the XeCl Laser Discharge," Journal of Physics D (Applied Physics), Vol. 28, No. 6, p. 1083-93, June 1995.

10. Itakura, Y., Nodomi, R., Kakimoto, M., "High Beam Quality ArF Excimer Lasers," Review of Laser Engineering, Vol. 20, No. 2, p. 75-85, Feb. 1992.

11. Boichenko, A.M., Derzhiev, V.I., Zhidkov, A.G., Yakovlenko, S.I., "Kinetic Model of an ArF Laser," Soviet Journal of Quantum Electronics, Vol. 22, No. 5, p. 444-8, May 1992.

12. Abele, C.C., Bunis, J.L., Caudle, G.F., Klauminzer, G., "Specifying Excimer Beam Uniformity," Proceedings of the SPIE – the International Society for Optical Engineering, Vol. 1834, p. 123-8, 1993.

13. Chumak, G.M., Nazarov, V.V., Smirnova, N.M., "Experimental Investigation of Factors which Restrict the Gas Lifetime in Excimer KrF Laser," Proceedings of the SPIE – The International Society for Optical Engineering, Vol. 1835, p. 203-12, 1993.

14. Mossavi, K., Hofmann, T., Tittel, F.K., "Ultrahigh-Brightness, Femtosecond ArF Excimer Laser System," Applied Physics Letters, Vol. 62, No. 11, p. 1203-5, March 1993.

15. Matsumoto, R., Katto, M., Okuda, M., Kurosawa, K., Sasaki, W., "Beam Profile Measurement of an Argon Excimer Laser with Imaging Plates," Review of Laser Engineering, Vol. 20, No. 9, p. 738-45, Sept. 1992.

16. Musso, M., Windholz, L., Fuso, F., Allegrnini, M., "Pulsed-Ultraviolet-Induced Chemiluminescence of NaCd and NaHg Excimers," Journal of Chemical Physics, Vol. 97, No. 9, p. 7017-18, Nov. 1992.

17. Hueber, J.M., Fontaine, B.L., Bernard, N., Forestier, B.M., Sentis, M.L., Delaporte, P.C., "Long Pulse KrCl Excimer Laser at 222 nm," Applied Physics Letters Vol. 61, No. 19, p. 2269-71, Nov. 1992.

18. Stevens, R.E., Kung, C.Y., Kittrell, C., Kinsey, J.L., "A simple and efficient external gas filtration and trapping system for excimer lasers," Review of Scientific Instruments, Vol. 65, No. 8, P. 2464-9, 1994.

19. Dong, H.K., Hong, C., Young, M.J., Sang, S.C., "Controllable pulse duration of a XeCl laser," Review of Scientific Instruments, Vol. 65, No. 12, P. 3634-8, 1995.

20. Rebhan, U., Patzel, R., Bucher, H. Powell, M., "High repetition rate lasers for advanced DUV exposure tools," SPIE, Vol. 2197, P. 920-6, 1994.

21. Kobayashi, Y., Ishihara,T., Nakarai, H. Itoh, N., Takahashi, T., Wakabayashi, O., Mizoguchi, H., Amada, Y., Fujimoto, J., Kowaka, M., Nozue, Y., "Recent advances of a KrF excimer laser on the plant's practical requirements," SPIE, Vol. 2197, P. 908-19, 1994.

3.10. UV SAFETY

22. Kunz, R.R., Horn, M.W., Bloomstein, T.M., Ehrlich, D.J., "Applications fo lasers in microelectronics and micromechanics," Applied Surface Science, Vol. 79-80, P. 12-24, 1994.

23. Hodgson, E., Boardman, A.D., Hua, Y., Wilson, A., "Optical measurement of excimer laser current pulses," Journal of Modern Optics, Vol. 41, No. 6, P. 1193-201, 1994.

24. Szatmari, S., "High-brightness ultraviolet excimer lasers," Applied Physics B (Lasers and Optics), Vol. B58, No. 3, P. 211-23, 1994.

25. Dragulinescu, D., Grigoriu, C., Tobos, L., "Compact excimer laser systems," SPIE, Vol. 1810, P. 410-13, 1994.

26. Frowein, H., Basting, D., "High power (500 W) excimer laser project," SPIE, Vol. 1810, P. 364-7, 1994.

27. Nagai, H., "Progress of excimer laser development in the AMMTRA project," SPIE, Vol. 1810, P. 348-55, 1994.

28. Letardi, T., Bollanti, S., DiLazzaro, P., Fang, H., Flora, L., Fu, S.F., Gerardino, A.M., Giordano, G., Lisi, N., Mezi, L., Schina, G., Torre, A., Zheng, C.E., "Excimer laser development and applications at the ENEA Frascati Centre," SPIE, 1994.

29. Patzel, R., Endert, H., "Excimer laser-a reliable tool for microprocessing and surface treatment," AIP Conference Proceedings, No. 288, P. 613-18, 1994.

30. Kowaka, M., Kobayashi, Y., Wakabayashi, O., Itah, N., Fujimoto, J., Ishihara, T., Makarai, H., Mizoguchi, H., Amada, Y., Nozue, Y., "Stability of krypton fluoride laser in real stepper mode operation," SPIE, Vol. 1927, Pt. 1, P. 241-51, Vol. 1, 1994.

31. Uteza, O.P., Sentis, M.L., Delaporte, P.C., Forestier, B.M., Fontaine, B.L., "Laser beam quality of high pulse repetition frequency excimer lasers," Optics Communications, Vol. 102, No. 5-6, P. 523-31, 1993.

32. Ross, I.N., Bialolenker, G., Evans, J., Hill, K., Hirst, G.J., Hooker, C.J., Houliston, J.R., Key, M.H., Lister, J.M., New, G., Sibbett, W., Shaw, M.J., Szatmari, S., Wilson, D, "Development of high-brightness KrF lasers using both Raman and ultra-short pulse CPA," SPIE, 1994.

Chapter 4

UV Materials Research

4.1 Introduction

Ultraviolet laser light is used in a number of specific applications and technology "niche" areas, several of which are the topics of individual chapters of this text: lithography, medicine, micromachining, cleaning, and semiconductor processing. Most of these applications were developed as a result of intense work in materials research, the subject of this chapter.

Materials research encompasses a wide range of material types and chemistry/light interactions. Researchers in UV technology prepare and run experiments that combine UV light, experimental materials (solids, films), various types of energy (heat, sound, magnetic, pressure), and a variety of controlled chemical (gas, liquid, solids) environments to observe and report on interactions, mechanisms, and phenomena. Data from work of this nature form the basis of emerging applications.

In this chapter we will discuss areas of UV material research including analytical techniques, photopolymer research, 157-nm laser technology, and ablation of PTFE. As useful applications are identified and pass the difficult test of being economically viable, they emerge from the research lab and move to the advanced development stage, where process optimization studies are used to define a working set of conditions for cost-efficient manufacturing.

4.2 Laser Mass Spectral Analysis

Laser ablation provides a means for a considerable number of diagnostic applications. The ablation reaction or phenomenon can provide a wide range of physical regimes ranging from low fluence thermal evaporation to UV plasma generation and high intensity inertial confinement fusion. The main advantage the UV laser provides for spectral analysis is high spatial resolution, sensitivity in the ppb (femtogram/attogram) range, athermal process at the surface, with little heating in the bulk of the substrate, and pinpoint focusing of the beam to allow analysis of very small sample sizes. Figure 4.1 shows a typical schematic for experiments using UV laser ablation (1).

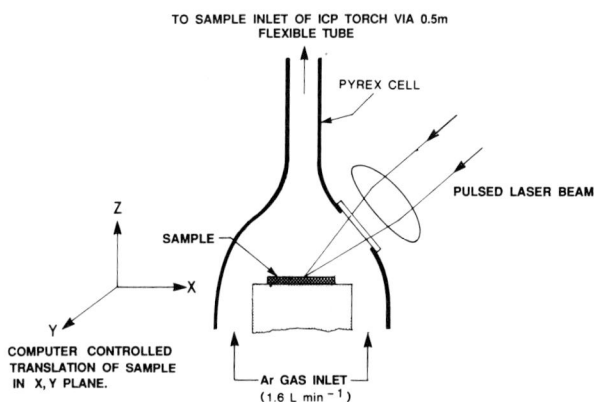

Figure 4.1: Schematic of UV Laser Ionization Mass Spectrometer

UV laser ablation produces many species, including particulates, ground-state atoms, excited atoms, and ions. These species are utilized for a number of elemental analysis techniques including atomic absorption (AA), laser ionization mass spectrometry (LIMS), atomic emission spectroscopy (AES), mass spectrometry (MS), laser-induced breakdown spectroscopy (LIBS), time of flight (TOF) analysis, mass analysis, and fourier transform ion cyclotron resonance mass spectrometry (FTICR).

Laser ablation is well suited for many types of analysis because samples can be used without being metallized or potted or prepared in any special way. Ablation can also be used to fragment metals, plastics, ceramics, glasses, and most any other kind of material. The spatial resolution is very high using a focused UV laser beam, permitting small areas of a sample to be used, or many individual sites to be tested from a single sample.

4.2. LASER MASS SPECTRAL ANALYSIS

Sample atomization, ionization, or excitation can be separated from sample vaporization or ablation, allowing individual and separated analyses to be run. For example, laser ablation can be combined with several of the other analytical methods such as AA, inductively coupled plasma (ICP), atomic emission spectroscopy (AES), or microwave-induced plasma (MIP) (2).

Solid sample introduction, using ICP-MS, has advantages for direct solid analysis compared to the more conventional methods such as x-ray fluorescence, spark, or arc AES. Benefits of UV laser ablation include analysis of polymers and other low density nonconductors, high detection limits, and little to no sample preparation. Excimer lasers, as well as quadrupled YAG or long wavelength lasers, are used as the source for initiating ablation for these analytical methods. Long wavelength laser mass specification analysis is not new, but the use of UV laser sources to drive reactions has emerged only as new UV optics and lasers have matured into commerical products.

4.2.1 MALDI Technique

UV mass spectral analysis can be accomplished in several ways, depending on the species to be analyzed. In the case of low energy UV analysis the main mechanism is desorption, where the removal of species such as large ions or individual molecules is desired. The sample is typically placed in a custom sample matrix and is irradiated with relatively low energy pulses. This method is called matrix-assisted laser desorption/ionization (MALDI). The combination of a special sample matrix and low UV energy permits extraction of individual, large molecules and molecule-sized species such as ion clusters (3).

4.2.2 UV Laser Microprobe Analysis

UV laser microprobe analysis is performed with a laser ionization mass spectrometer (LIMS). This method relies on the use of relatively high energy laser pulses and high fluence (125 mW/cm^2). High energy laser pulses and fluence will, in many cases, cause fragmentation, ionization, or vaporization of the sample (4). Laser microprobe analysis is useful for studying many of the high density thin films used in semiconductor device fabrication, such as tantalum silicide and polysilicon.

UV laser microprobe analysis at slightly lower (5-25 mW/cm^2) energy levels is also useful for a regime where the secondary processes of electron

CHAPTER 4. UV MATERIALS RESEARCH

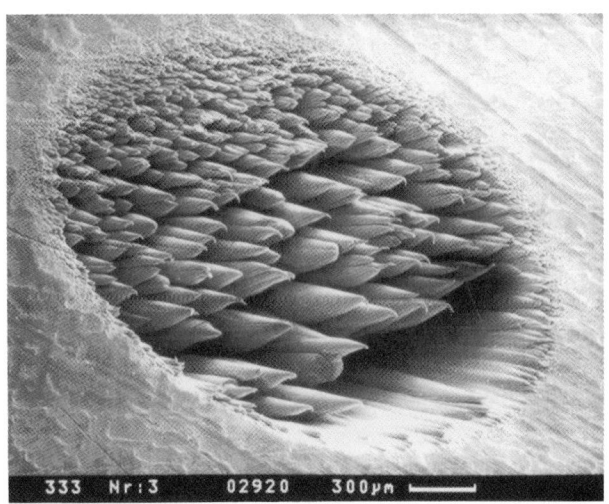

Figure 4.2: Ablation of PTFE Using a 157-nm Fluence

impact, photoionization, and excited plasma generation play a dominant role in isolating the sample and its constituents. The lower energy regime simply allows for fragmentation or ionization of larger "chunks" of the sample with the assistance of a second laser beam after initial volatilization with the primary beam. For example, high energy ablation of relatively low density polymers rarely leaves complete molecules intact; lower energy levels permit a much larger number of intact large molecules to survive (5).

4.3 157-nm Fluorine Laser Processing

Fluorine lasers emit light in the vacuum ultraviolet at 157.5 and 157.6 nm and some light in the red spectral region. Similar to the other excimer lasers discussed in this text, fluorine lasers are excited by high pressure discharges. Other high energy sources, such as electron beams, can be used but are not practical.

The commercialization of the 157-nm laser has permitted a new capability for deep-UV materials research. Some materials have low absorption above 160 nm and therefore do not respond to exposure with the more common 193- and 248-nm excimer laser wavelengths. Teflon, for example, has very low absorption above 160 nm, but absorbs well at the 157-nm fluorine laser line. Figure 4.2 shows the result of a 157-nm ablation of PTFE using a fluence of 0.3 J/cm^2. The ablation rate was 0.15 μm per pulse; at 1.2 J/cm^2 it increased to 0.3 μm per pulse (8).

Efficient coupling of light at very short UV wavelengths provides the means for structuring materials with very high resolution and with greater fracturing in the ablation process than at longer wavelengths, due mainly to the higher photon energy available at 157 nm compared to 193- and 248-nm wavelengths. In optical lithography, greater patterning resolution is possible with the deep-UV wavelengths. This is making excimers a production tool for IC manufacturing.

The vacuum ultraviolet is especially important in UV technology for the high photon energy and the many reactions which can be driven as a result. The quantum energy of a 157-nm photon is 7.9 eV. This is sufficient energy to excite materials to high electronic states and, in many cases, to cause ionization of atoms and molecules with a single pulse or short pulse train (6).

Fluorine lasers require high pressure, and the absorption of 157-nm radiation by many molecules increases the sensitivity of the laser to impurities in the discharge chamber and laser gas impurities. In an operating laser, secondary reactions between the gas mixture and corrodible materials in the discharge chamber also introduce impurities which absorb the 157-nm photons. Liquid nitrogen gas cryogenic processing of the laser gas is generally needed for 157-nm lasers to keep the output high and stable. Cryogenic cooling of the liquid nitrogen to -210° C will provide the necessary freezing out of the impurities caused by electrode ablation and other gas interactions and reactions with some of the materials in the discharge chamber. More inert materials, such as refractory ceramics, are being used in excimer lasers to reduce this problem.

4.4 Production of X-Rays with 248 nm Energy

The focusing of 248-nm excimer laser energy into a variety of different target materials is a useful and efficient method for x-ray production. To produce the x-rays, a KrF laser with a 200-J, 50-nsec pulse is delivered in a 265-mm diameter beam (annular), as show in Figure 4.3. A fused silica lens focuses the beam, followed by a silica shield to catch debris from ablation reactions. The excimer laser beam strikes the target with relatively low divergence to

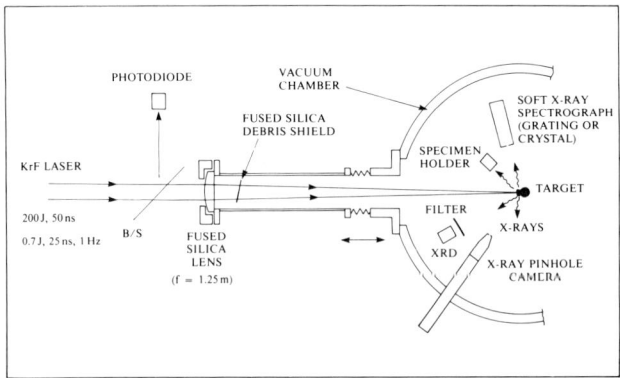

Figure 4.3: KrF Laser Plasma XUV Source

produce a small spot. Divergence is about 20μrad.

The target in the vacuum (10^{-5}) chamber can be gold, aluminum, carbon, or other materials. Soft x-rays from aluminum targets can be produced with 0.2% conversion efficiency. Carbon and gold targets were used to produce 280-eV x-ray photons, used for contact x-ray microscopy of living biological cells. A specimen chamber for this type of work is shown in Figure 4.4.

4.5 Excimer Laser-Induced Fluorescence

Laser diagnostics is widely used to map a variety of complex chemical processes, especially applications involving complex reaction kinetics. In the past, lasers have been used for combustion analysis in stagnant flames and gas flows. More recently, excimer lasers have been used to map the concentrations of transient OH molecules in an application to provide a thorough analysis of the combustion process. In excimer laser-induced fluorescence, OH molecule concentration is mapped as a snapshot in the time frame when the flames are propagating in a gas/air mixture.

Figure 4.5 shows a sequence where the OH molecule density (dark areas are lowest concentration) in a combustion chamber where the piston movement and gas/air parameters induce turbulence and density changes in OH molecule concentration. This detailed molecular analysis of the combustion

4.6. TUNABLE UV LASER

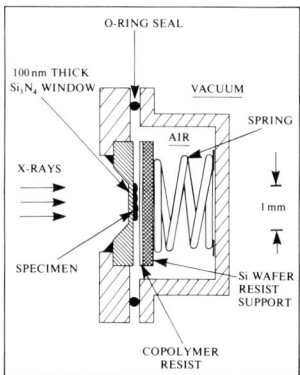

Figure 4.4: Specimen Chamber for Biological X-Ray Microscopy

process permits optimization for fuel economy and engine reliability not before possible. The laser pulse energy was 350 mJ and a pulse duration of 17nsec. Note the concentration of OH molecules along the flame front.

4.6 Tunable UV Laser

In UV spectroscopy there are a variety of laser requirements, including the need for wavelength tuning, spectral purity, brightness, and spectral narrowing. A tunable UV laser system illustrated in Figure 4.6, with 60 MHz linewidth and 10 mJ pulse energy is composed on commercial products. The system uses a continuous ring dye laser pumped by a 75-W, 514-nm argon laser. The output is about 300 mW at 730 nm.

The output of the dye laser is sent into a three stage amplifier which is fed by an excimer laser. The dye used in the amplifier is rhodamine dissolved in methanol.

The splitting of the pulses thorough the amplifiers will optimize the beam for frequency doubling. The last amplifier (for the UV) incorporates a DMY dye dissolved in dioxane. The tuning of the system is accomplished by tuning the ring laser. A typical pulse is about 20 nsec (FWHM), and using fresh dyes and fresh gas, pulse energies as high as 20 mJ were obtained.

102   CHAPTER 4. UV MATERIALS RESEARCH

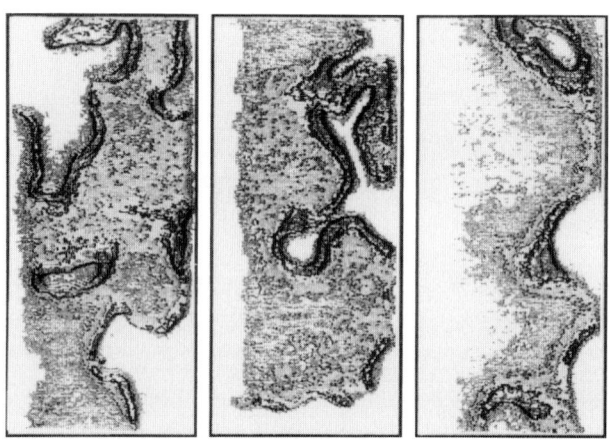

Figure 4.5: LIF Mapping of OH Molecules Excited by 308-nm Laser

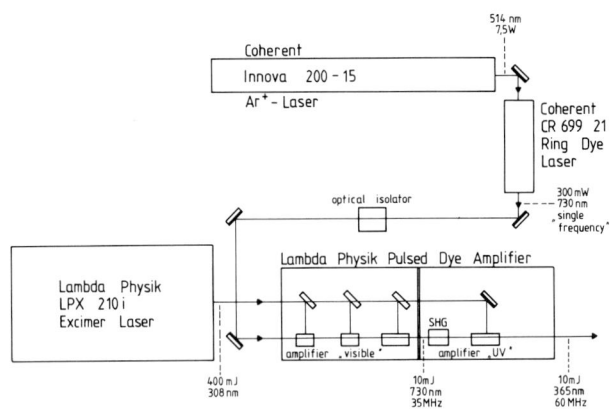

Figure 4.6: Tunable UV Laser System

4.7. LASER CHEMICAL VAPOR DEPOSITION   103

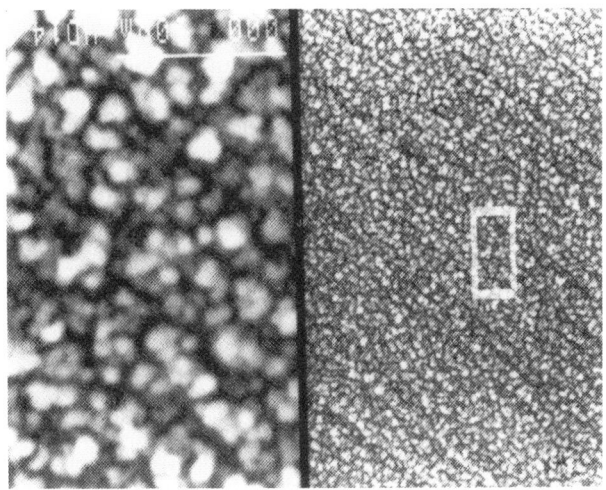

Figure 4.7: Aluminum Deposits Using Twin Processing

4.7 Laser Chemical Vapor Deposition

Photochemical reactions induced by lasers are widely used for a deposition technology called laser chemical vapor deposition (LCVD). The use of excimer lasers in film deposition is multipurpose. UV energy is useful in preparing the surface before deposition, providing energy to drive the deposition reaction, and in treating the final deposited film (hardening, alloying, and annealing). In this section we will focus on twin beam processing with excimer lasers where one beam is used for fragmenting the precursor compound and the other is used either in pre- or post-treatment steps, or in the reaction itself. Pretreatment involves heating and photonucleation, while the CVD reaction requires that the laser energy induce gas phase reactions or photo fragmentation of a compound. The post-treatment step is to surface alloy the film or cure the deposit to promote increased adhesion.

Figure 4.7 shows aluminum deposits made using twin processing from trimethyl-aluminum (TMA) in a nickel alloy substrate. Note the grain size of about 2μm.

4.8 Surface Analysis by Laser Ionization

UV laser radiation is used in the SALI technique to eject material from a surface and ionize neutral molecules. The ions pass through a time-of-flight mass spectrometer and are counted on a microchannel plate detector. It has been found that SALI techniques can complement conventional secondary ion mass spectrometry (SIMS).

Figure 4.8 shows a comparison of GaAs surface. The excimer beam used was a 248-nm laser passed millimeters above the surface where it intersected a cloud of desorbed neutral atoms and molecules released by argon ions. The goal is to ionize the neutrals by multiphoton ionization (MPI) nonresonantly. While the data from SIMS and SALI complement each other, the SALI data are more quantitative since they do not rely on secondary ions.

4.9 Photopolymer Research

Organic coatings know as "photopolymers" or "photoresists" are a widely used class of materials that serve as imagable masks in etching, electroplating, and electroforming processes. Photoresists play a critical role by defining the patterns used to form integrated circuits, printed circuits, thin film circuits, and electroformed and micromachined parts. Photoresist is also used to make compact disc masters and stampers, thin film heads for computers, and flat panel displays (LCDs).

Most of the applications for photoresist require high resolution, high fidelity (1:1 transfer of the mask dimensions), or both.

Deep UV-sensitive photoresists are being developed to provide increased imaging resolution. The shift to a shorter exposure wavelength is the most direct and cost effective means to higher resolution. A lithography research effort is also being made in deep-UV optics to produce high NA lenses to image the deep-UV resists.

Image transfer with deep-UV resists is governed by the light source, the mask pattern, the imaging optical system, and photoresist contrast and resolution properties. Figure 4.9 shows the mask pattern and its transform into a signal of varying intensity in the resist. The resist response to a projected aerial image is critical in determining the ability to faithfully reproduce the mask dimensions.

Interference effects also play a role in the resist response during exposure. Incident light is reflected back from the substrate as a function of the

4.9. PHOTOPOLYMER RESEARCH

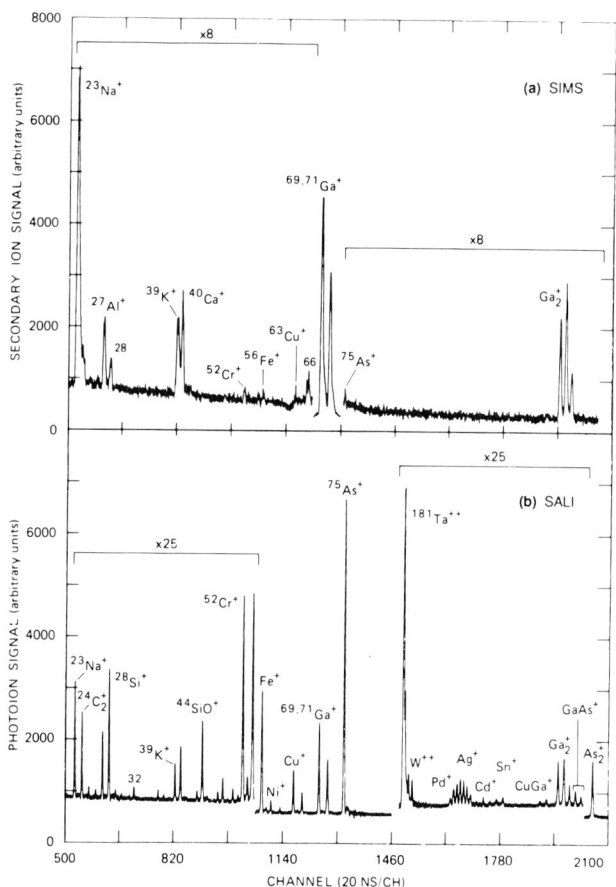

Figure 4.8: Comparison of SIMS and SALI Analysis

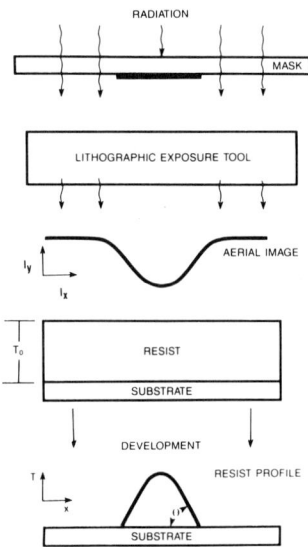

Figure 4.9: Mask Pattern and Transform to Resist Image

optical thickness, the surface reflectively, and surface topography. Reflected light from the substrate can cause standing waves or produce areas of either increased or decreased exposure, depending on the substrate. An example of these various conditions is shown in the schematic example of exposure mechanics in Figure 4.10

The required exposure dose to clear or permit complete development (positive-tone resist) of the resist varies considerably in the example. Standing waves also impact the exposed and developed image parameters by producing alternating nodes of intensity maxima and minima. The intensity profile is reproduced in the resist sidewall and causes linewidth variations.

The reflectivity of the substrate is also impacted by the exposure wavelength. For example, the reflectivity of tungsten silicides at deep-UV wavelengths is about 50%; silicon is 65% reflective; aluminum silicon-copper is 70% at 193 nm and 85% at 248 nm, and aluminum-silicon is 85% at 193 nm and 90% at 248 nm.

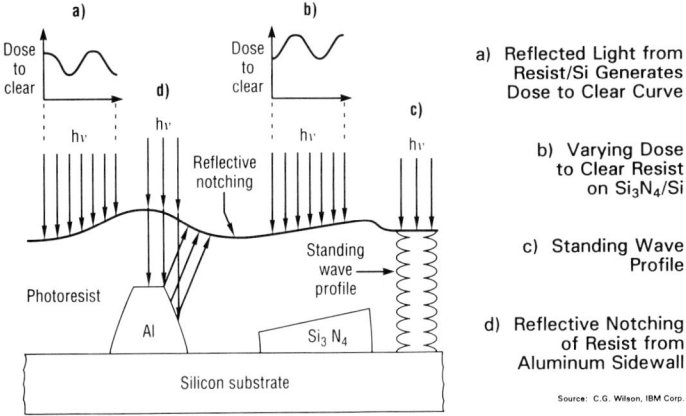

Figure 4.10: Interference Effects in Photoresist

4.10 Laser Treatment of Organosilicons

The effects of 193-nm laser exposure on films and coatings has been measured by many researchers. In one study, a 193-nm dose proportional transformation of organosilicons tetramethylsilane (TMS) and hexamethyldisilizane (HMDS) into quartz-like compounds was observed. Figure 4.11 shows the infrared absorption spectrum of a 100-nm layer as a function of the number of pulses. The data show absorption increases in the spectral domain of the Si-O nitrational bond.

Figure 4.12 shows the effect of an argon fluoride laser in UV transmission of a 50-nm layer of HMOS; data show transmission below 220 nm increasing with dose.

4.11 UV Laser Cleaning

UV laser energy is being increasingly used as a means to drive reactions on surfaces which either remove unwanted material from the surface or alter the surface state without depositing significant thermal energy at the surface or below in the bulk of the substrate. The research in this area is directed at photolytic as opposed to photothermal mechanisms. Photolytic bond

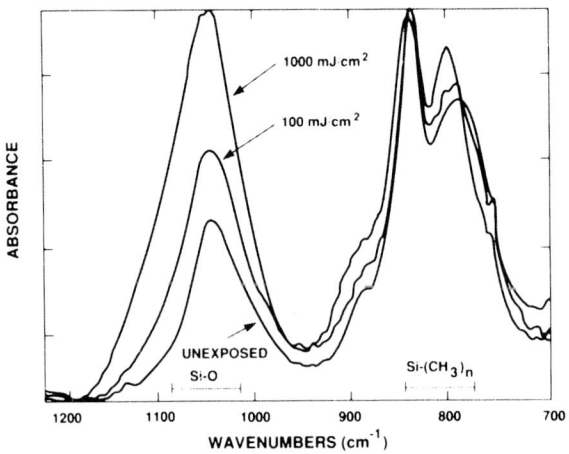

Figure 4.11: 193-nm Dose Effect on IR Spectrum of TMS

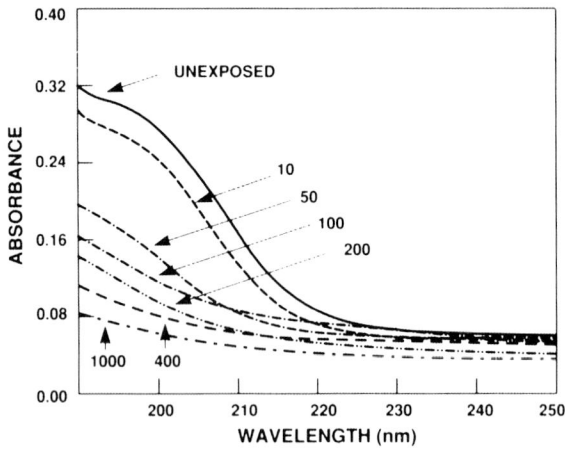

Figure 4.12: 193-nm Dose Effect on UV Transmission of HMDS

4.11. UV LASER CLEANING

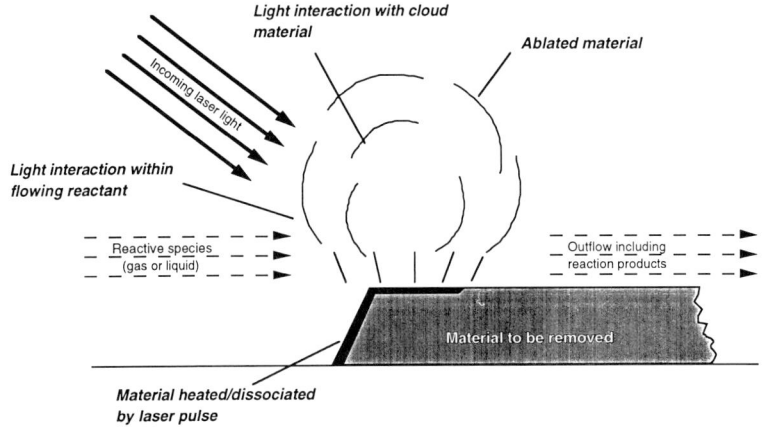

Figure 4.13: Schematic of a UV Laser and Reactive Gas Cleaning

breaking is one mechanism for athermal surface cleaning wherein the inherent quantum mechanical properties of the UV source are efficiently utilized. Efficient photon coupling onto a material for photolytic bond breaking is driven by fundamental constants such as the wavelength and photon energy.

The principal mechanisms for UV laser-assisted gas cleaning are complex since they involve ablation. Figure 4.13 shows a schematic of a focused UV laser blade sweeping in the presence of a reactive gas.

Forces that bond materials to surfaces include van der Waals, dipole-dipole, hydrogen, electrostatic, and ionic. The covalent bonds are the strongest bonds. Particles that are liberated from surfaces by simple photolytic or photothermal bond-breaking reactions need to be removed from the surface. Some researchers attempt to do this by flowing inert gas across the surface in the presence of UV radiation which is used to break bonds (9).

UV-assisted surface treatment has been investigated by other researchers who use a reactive gas in the presence of UV radiation; the UV light in this case not only breaks bonds, but provides excitation of certain chemi-

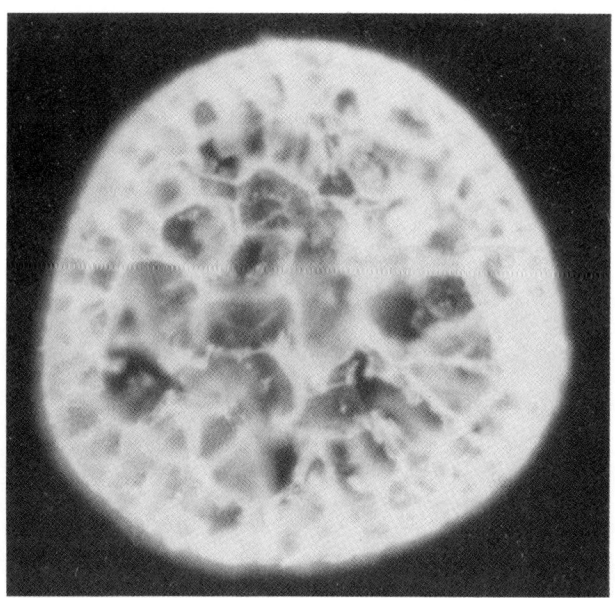

Figure 4.14: Cross Section of a Root Ablated at 193 nm

cal species in gases and the excited gas molecules convert the ablated by-products into gaseous by-products so that only gaseous by-products are produced, eliminating particulates.

4.12 Cross-Sectioning Delicate Structures

Excimer laser ablation is applied to the cross-sectioning of delicate structures in forestry research, where small roots grown under experimental conditions must be analyzed internally. Figure 4.14 shows the cross section of a small root specimen cut with a 193-nm excimer laser.

References

1. "Laser Ablation," Springer Series in Materials Science, Vol 28, Springer-Verlag.

2. Arrowsmith, P., "Laser Ablation of Solids for Elemental Analysis by Inductively Coupled Plasma Mass Spectrometry," Analytical Chemistry, 1987, 59, 1437.

3. Miller, J.C., Ed., "Laser Ablation," Springer Series in Materials Science, Vol 28, Springer-Verlag.

4. Arrowsmith, P., "Laser Ablation of Solids for Elemental Analysis by Inductively Coupled Plasma Mass Spectrometry," Analytical Chemistry," 1987, 59, 1437.

5. Sappey, A.D., Nugan, N.S., "Diagnostic Studies of Laser Ablation for Chemical Analysis," Laser Ablation, Ed., J.C. Miller, Springer,-Verlag, P. 157-159, 1994.

6. Ledingham, K.W., "Analytical Proceedings" London 28, 413, 1991.

7. Nagan, N.S., Estler, P.C., Miller, C.M., "Analytical Chemistry, 57, 2441, 1985.

8. Lambda Physik Technical Publication, "Lambda Highlights," Number 25, October 1990.

9. Wilson, C.G., IBM Corporation, Technical Publication from course on "Deep UV Photoresist Technology."

10. Engelsberg, A., "Laser Assisted Cleaning Proves Promising," Cleaning Alternatives, P. 35-42, May 1995.

11. U.S. Patent Numbers 25,024,968/5,099,557, Engelsberg, A.,

Glossary of Terms

AES Atomic emission spectroscopy.

FTICR Fourier transform in cyclotron resonance mass spectrometry.

ICP Inductively coupled plasma.

Intrinsic Photoemission The photoemission that would occur if a crystal were free of any impurities and its crystal planes were without any defects or dislocations.

LCD Liquid crystal displays.

LIBS Laser-induced breakdown spectroscopy.

LIMS Laser ionization mass spectrometry.

LCVD Laser chemical vapor desposition.

MALDI Matrix-assisted laser desorption/ionization.

MIP Microwave-induced plasma.

MS Mass spectrometry.

SALI Surface analysis by laser ionization.

TOF Time of flight.

Reading Bibliography

1. Comita, P.B., Farkas, J., Yang, B., Chuang, Y.H., O'Connor, J., Liu, K.C., Cohen, M.G., "Pulsed Ultraviolet Laser Deposition of SiO_2 Films at 248nm," Applied Physics Letters, Vol. 66, No. 12, p. 1463-5, 1995.

2. Askar'yan, G.A., Korolev, M.G., Yurkin, A.V., Yakushkin, K.L., "Cumulation, Collisions, and Flow Around Fas Gas Jets in Laser Ablation of Surfaces an in Explosions of Thin Films of Matter. Practical Applications," Quantum Electronics, Vol. 25, No. 4, p. 289-90, April 1995.

3. Dubowski, J.J., "Pulsed Laser Ablation: A Method for Deposition and Processing of Semiconductors at an Atomic Level," Materials Science Forum, Vol. 173-174, p. 73-80, 1995.

4. Pedraza, A.J., DeSilva, M.J., Kumar, R.A., Lowndes, D.H., "The Adhesion of Metallic Films to Laser-Irradiated Alumina," Journal of Applied Physics, Vol. 77, No. 10, p. 5176-9, May 1995.

5. Bublitz, J., Dickenhausen, M., Gratz, M., Todt, S., Schade, W., "Fiber-Optic Laser-Induced Fluorescence Probe for the Detection of Environmental Pollutants," Applied Optics, Vol. 34, No. 18, p. 3233-33, June 1995.

4.12. CROSS-SECTIONING DELICATE STRUCTURES

6. Zuber, A., Kaiser, N., Stehle, J.L., "Variable-Angle Spectroscopic Ellipsometry for Deep-UV Characterization of Dielectric Coatings," Thin Solid Films, Vol. 261, No. 1-2, p. 37-43, June 1995.

7. Srinivassa, R., "Etching Polymer Films with Ultraviolet Laser Pulses of Long Duration," Journal of Applied Physics, Vol. 72, No. 4, p. 1651-3, Aug. 1992.

8. Greer, J.A., "Comparison of Large-Area Pulsed-Laser Deposition Approaches," Proceedings of the SPIE – The International Society for Optical Engineering, Vol. 1835, p. 21-30, 1993.

9. Kuper, S., Branon, J., "KrF-Laser Ablation of Polyurethane," Applied Physics A (Solids and Surfaces), Vol. A57, No. 3, p. 255-9, Sept. 1993.

10. Montagne, J.E., Sarnet, T., Prat, C., Inglesakis, G., Autric, M., "High-Intensity KrF Excimer Laser Processing of Metal Surfaces," Applied Surface Science, Vol. 69, No. 1-4, p. 108-14, May 1993.

11. Gorbunov, A.A., Konov, V.I., "Excimer Laser Ablation for Thin Film Deposition," Laser Ablation of Electronic Materials, Basic Mechanisms and Applications, p. 377-84.

12. Hongping, G., Qihong, L., "Study of Gas-stream-Assisted Laser Ablation of Copper," Thin Solid Films, Vol. 218, No. 1-2, p. 274-6, Oct. 1992.

13. Yabe, A., Niino, H., "Surface Modification of Polymers with Excimer Lasers and its Applications," Laser Ablation of Electronic Materials, Basic Mechanisms and Applications, p. 199-212.

14. Bergmann, H.W., Schubert, E., Schutte, K., "Surface Modifications Using Excimer Lasers: Fundamentals and Applications," Journal de Physique IV (Colloque), Vol. 1, No. C7, p. 7-12, Dec. 1991.

15. Mizrahi, V., Lemaire, P.J., Erdogan, T., Reed, W.A., DiGiovanni, D. J., Atkins, R.M., " Ultraviolet laser fabrication of ultrastrong optical fiber gratings and of germania-doped channel, waveguides," Applied Physics Letters, Vol. 63, No. 13, P. 1727-9, Feb. 1994.

16. Vega, F., Afonso, C.N., Ortega, C., Sieijka, J., "Kinetics of pulsed ultraviolet laser induced oxidation of c-ge: the role of optical coupling

and material losses," Journal of Applied Physics, Vol. 74, No. 2, P. 963-8, 1993.

17. Benattar, R., "Laser produced plasmas in the x-uv region: application to calibration," Ann. Phys. (France), Annales de Physique, Vol. 17 colloq, No. 1, P. 39-45, 1993.

18. Gruzinskii, V.V., Senyuk, M.A., Neyra Bueno, O.L., Afanasiadi, L.Sh., "Laser ability and luminescent spectral properties of compounds with alternating phenyl, oxazole, oxadiazole, furyl, and ethylene units," Zhurnal Prikladnoi Spektroskopu, Vol. 57, No. 1-2, P. 82-6, 1993.

19. Fontaine, B., Delaporte, Ph., Sentis, M., Forestier, B., "Recent progress by the cnrs imfm-um34 new lasers group in the field of pulsed high power coherent and incoherent UV-VUV sources," Ann. Phys. (France), Annales de Physique, Vol;. 17 colloq, No. 1, P. 179-80, 1993.

20. Katto, M, Matsumoto, R., Kurosawa, K., Sasaki, W., Takigawa, Y., Okuda, M., "Laser beam profiler in the vacuum ultraviolet spectral range using photostimulable phosphor," Review of Scientific Instruments, Vol. 64, No. 2, P. 319-24, 1993.

21. Watanabe, A., Imai, Y., Osato, K., Mukaida, M., Sugawara, K., Takeo, H., Kameyama, T., Fukuda, K., "Effect of operating conditions of KrF excimer laser on crystallinity of deposits in LCVD from $Mo(CO)_6$," Denki Kagaku, Vol. 60, No. 11, P. 1009-11, 1993.

22. McCoustra, M.R.S., Hippler, M., Pfab, J., "The 355 nm laser photolysis of jet-cooled methyl nitrite (CH_3ONO). Internal energy distributions of the NO fragment," Checmical Physics Letters, Vol. 200, No. 5, P. 451-8, 1993.

23. Egitto, F.D., Davis, C.R., "Dopant-induced excimer laser ablation of poly (tetrafluorethylene). II. Effect of dopant concentration," Applied Physics B (Photophysics and Laser Chemistry), Vol. B55, No. 6, P. 488-93, 1993.

24. Fogarassy, E., Lazare, S., "Ultraviolet laser interactions with polymer surfaces in the microsecond regime: The photokinetic effect," Laser Ablation of Electronic Materials. Basic Mechanisms and Applications, p. xi+394, 191-7, 1993.

4.12. CROSS-SECTIONING DELICATE STRUCTURES

25. Feldman, L.A., "Mechanical testing of laser-cured photopolymers for multiple-layer structures," Journal of Materials Science Letters, Vol. 11, No. 18, P. 1231-3, 1992.

26. Kurosawa, K., Sasaki, W., Takigawa, Y., "Surface modification by superdry processes with vacuum ultraviolet lasers," Review of Laser Engineering, Vol. 20, No. 1, P. 11-19, 1992.

27. Liu, J., Liu, Z., Zhang, M., Zhu, S., Wu, Z., "Epitaxial growth of optical waveguiding Lithium-Tantalum-Oxide films by excimer laser ablation," Journal of Physics Condensed Matter, Vol. 6, No. 28, P. 5409-14, July 1994.

28. Kennedy, E.T., Costello, J.T., Mosnier, J.P., Cafolla, A.A., Collins, M., Kiernan, K., Koble, U., Sayyad, M.H., Shaw, M., Sonntag, B.F., Barchewitz, R., "Extreme-ultraviolet studies with laser-produced plasmas," Optical Engineering, Vol. 33, No. 12, P. 3984-92, 1995.

29. Affatigato, M., Tang, K., Haglund, R.F., Jr., Chen, C.H.,"Ultraviolet-laser-induced desorption of atoms, ions, and molecules from lithium niobate," Applied Physics Letters, Vol. 65, No. 14, P. 1751-3, 1994.

30. Song, M.B., Suguri, M., Fukutani, K., Komori, F., Murata, "Laser-induced desorption and etching at surfaces," Appl. Surf. Sci. (Netherlands,), Applied Surface Science, Vol. 79-80, P. 25- 33, 1994.

31. Hayase, R., Onishi, Y., Niki, H., Oyasato, N., Hayase, S., "Chemically amplified resists using 1,2-naphtoquinone diazide-4-sulfonates as photoacid generators," Journal of Electrochemical Society, Vol. 141, No. 11, P. 3141-5, 1995.

32. Nakata, Y., Sasaki, Y., Okada, T., Maeda, M., "Radical beams produced by laser ablation and their application," Japan Journal of Applied Physics, 1 Regular Papers, & Short Notes, Vol. 33, No. 7B, P. 4316-19, 1995.

33. Askar'yan, G.A., Korolev, M.G., Korchagina, E.G., Yakushkin, K.L., "Effect of intense UV flashes from laser discharges in gases: observation of a fast-rising photodissociation halp leading to a shock wave," Pis'ma v Zurnal Eksperimental'noi i TeoreticheskoiFiziki, Vol. 60, No. 1, P. 11-15, 1994.

34. Searles, S.K., Tucker, J.E., Wexler, B.L., Masters, M.F., "VUV emission from discharge-pumped Ar supersonic jets under high-excitation conditions," IEEE Journal of Quantum Electronics, Vol. 30, No. 9, P. 2141-6, 1994.

35. Ardakani, H.K., "Electrical and optical properties of in situ 'hydrogen-reduced' titanium dioxide thin films deposited by pulsed excimer laser ablation," Thin Solid Films, Vol. 248, No. 2, P. 234-9, 1994.

36. Dawson, P., Cairns, G, "Surface plasmon enchanced laser ablation of thin metal films," J. Mod. Opt, (UK), Journal of Modern Optics, Vol. 41, No. 6, P. 1287-94, 1994.

37. Sterkenburgh, T.R., Franke, H., Frank, W., Garen, W., Becker, M.G., "Patterning polymer surfaces with laser radiation for fiber chip coupling," Proc. SPIE - Int. Soc. Opt. Eng. (USA), Proceedings of the SPIE - The International Society for Optical Engineering, Vol. 2042, P. 455-61, 1994.

38. Fujinami, M., Chilton, N.B., "Study on excimer laser induced defects in SiO_2 films on Si by variable-energy positron annihilation spectroscopy," Applied Physics Letters, Vol. 64, No. 21, P. 2806-8, 1994.

39. Tsuboi, Y., Fukumura, H., Masuhara, H., "Nanosecond imaging study on laser ablation of liquid benzene," Applied Physics Letters, Vol. 64, No. 20, P. 2745-7, 1994.

40. Mihailov, S.J., Gower, M.C., "Recording of efficient high-order Bragg reflectors in optical fibres by mask image projection and single pulse exposure with an excimer laser," Electronics Letters, Vol. 30, No. 9, P. 707-9, 1994.

41. Kinsman, G., Duley, W.W., "Treatment of metal surfaces with excimer laser radiation for radiative applications," Applied Optics, Vol. 32, No. 36, P. 7462-70, 1994.

42. Sugioka, K., Wada, S., Tsunemi, A., Sakai, T., Takai, H., Moriwaki, H., Nakamura, A., Tashiro, H., Toyoda, K., "Micropatterning of quartz substrates by multi-wavelength vacuum-ultraviolet laser ablation," Japan Journal of Appllied Physics, Part 1, Regular Papers, & Short Notes, Vol. 32, No. 12B, P. 6185-9, 1994.

4.12. CROSS-SECTIONING DELICATE STRUCTURES

43. Pedraza, A.J., Park, J.W., DeSilva, M.J., Lowndes, D.H., Braski, D.N., Meyer, H.M., III, "Mechanisms of laser activation of dielectric materials," AIP Conference Proceedings (USA), No. 288, P. 329-34, 1994.

44. Jarvis, T.R., Nastasi, M., Hirvonen, J.P., "Excimer laser surface modification: process and properties," Beam Solid Interactions: Fundamentals and Applications Symposium, P. xvii +913, 665-78, 1994.

45. Blanchet, G.B., Fincher, C.R., Jr., Jackson, C.L., Shah, S.I., Gardner, K.H., "Laser ablation and the production of polymer films," Science, Vol. 262, No. 5234, P. 719-21, 1994.

46. Ikuta, Koji, Hayashi, Michihiro, Matsuura, Toshihiko, Fujishiro, Hiroyuki, "Shape memory alloy thin film fabricated by laser ablation," Proceedings of the IEE Micro Electro Mechanical Systems, P. 355-360, 1994.

47. Xiong, Fulin, Hagerman, M.E., Zhou, H., Lundquist, P.M., Wong, G.K., Poeppelmeier, K.P., Ketterson, J.B., Chang, R.P.H., White, C.W., "Fabrication of potassium titanyl phosphate films by pulsed excimer laser ablation for nonlinear optical application," Proceedings of the 1993 Fall Meeting, 1994.

48. Sugioka, Koji, Wada, Satoshi, Tsunemi, Akira, Sakai, Toshiaki, Takai, Hiroshi, Moriwaki, Hiroki, Nakamura, Akira, Tashiro, Hideo, Toyoda, Koichi, "Micropatterning of quartz substrates by multi-wavelength vacuum-ultraviolet laser ablation," Japanese Journal of Applied Physics, Part 1: Regular Papers, Short Notes, Review Paper Vol. 32, No. 128, P. 6185-89, Dec. 1993.

49. McIntosh, J., Zervaki, A., Papadimitriou, K., Haidemenopoulos, G.N., Manousaki, A., Zergioti, G., Hontzopoulos, E., "Excimer laser used as a materials characterization tool: Sulphide inclusion printing in steel," Laser Chemistry, Vol. 13, No. 2, P. 121-28, 1993.

50. Nistor, L.C., VanLanduyt, J., Ralchenko, V.G., Kononenko, T.V., Obraztsova, E.D., Strelnitsky, V.E., "Direct observation of laser-induced crystallization of C:H films," Applied Physics A: Solids and Surfaces Vol. 58, No. 2, P. 137-44, Feb. 1994.

51. Johnston, C., Chalker, P.R., Buckley-Golder, I.M., Marsden, P.J., Williams, S.W., "Diamond device delineation via excimer laser patterning," Diamond and Related Materials, Vol. 2, No 5-7, Pt 2, P 829-34, Apr. 1993.

52. Bourdon, E.B.D., Kovarik, P., Prince, R.H., "Microcrystalline diamond phase by laser ablation of graphite," Diamond and Related Materials Vol. 2, No. 2-4, Pt 1, P. 425-31, Mar. 1993.

53. Gendler, Z., Rosen, A., Bamberger, M., Rotel, M., Zahavi, J., "Improvement of adhesive bonding strength in sealed anodized aluminum through excimer laser prebond treatment," Journal of Materials Science, Vol. 29, No. 6, P. 1521-26, Mar. 1994.

54. Blanchet, Graciela B., Fincher, C.R., Jr., Jackson, C.L., Sheh, S.I., Gardner, K.H., "Laser ablation and the production of polymer films," Science Vol. 262, No. 5134, P. 719-21, Oct. 1993.

55. Nowak, R., Metev, S., Sepold, G., Grosskopf, K. "Excimer laser processing of BK7 and BGG31 glasses," Glastechniache Berichte, Vol. 66, No. 9, P. 227-33, Sep. 1993.

56. Song, Wendong, An, Chengwu, L., Dongsheng, Fan, Yongchang, Li, Zaiguang, Huang, Jingqing, Zhang, HonghaiZhao, Xingrong, Mai, Zhihong, Zhou, Fangqiao, Guo, Qingweii, Li, Yuenan, Hua, Zhigiang, "Deposition of large area superconducting films by excimer laser scanning ablation"Wuli Xuebao/Acta Physics Sinica, Vol. 42, No. 10, P. 1674-79, Oct. 1993.

57. Anderson, C., "Photoablation of Dupont Kapton at energy desities in excess of 5J/sq.cm," Proceedings of SPIE, Vol. 1990, P. 415-421, 1993.

58. Phillips, H.M., Feurer, T., LeBlanc, S.P., Callahan, D.L., Sauerbrey, R., Bentley, J., "Excimer laser induced permanent electrical conductivity and nanostructures in polymers," Proceeding SPIE, Vol. 1856, P. 143-54, 1994.

59. Sorescu, M., Knobbe, E.T., "Effect of pulsed excimer laser irradiation on the magnetic anisotropy of iron-base and nickel-iron-base metallic glases," Journal of Materials Research, Vol. 8, No. 12, P. 3078-84, 1994.

4.12. CROSS-SECTIONING DELICATE STRUCTURES

60. Lowndes, D.H., Godbole, M.J., Jellison, G.E., Jr., Pedraza, A.J., "Ablation, surface activation, and electroless metallization of insulating materials by pulsed excimer laser irradiation," AIP Conference Proceedings, No. 222, P. 321-8, 1994.

61. Geiger, M., Lutz, N., Rebhan, T., Hutfless, J., "Quality control during excimer laser material processing," SPIE, Vol. 1810, P. 620-3, 1994.

62. Godbole, M.J., Lowndes, D.H., Pedraza, A.J., "Excimer laser ablation and activation of SiO_x and SiO_x-ceramic couples for electroless copper plating," Applied Physics Letters, Vol. 63, No. 25, P. 3449-51, 1994.

63. D'Couto, G.C., Babu, S.V., Egitto, F.D., Davis, C.R., "Excimer laser ablation of polyimide-doped poly (tetrafluorethylene) at 248 and 308 nm," Journal of Applied Physics, Vol. 74, No. 10, P. 5972-80, 1994.

64. Haedemonopoulos, G.N., Zervaki, A., Papdimitriou, K., Tsipas, D.N., McIntosh, J., Zergioti, G., Manousaki, A., Hontzopoulos, E., "Surface treatment of metals with excimer and Co_2 lasers," SPIE, Vol. 1810, P. 712-15, 1994.

65. Toyoda, K., Sugioka, K., "Surface hardening and improvement of corrosion resistance of SUS304 by KrF excimer laser irradiation in SiH_4 gas ambient," SPIE, Vol. 1810, P. 708-11, 1994.

66. Lazare, S., Benet, P., Bolle, M., DeDonato, P., Bernardy, E., "New surface modifications of polymer films with the excimer laser radiation," SPIE, Vol. 1810, P. 546-53, 1994.

67. Mihailov, S., Lazare, S., "Fabrication of refractive microlens arrays by excimer laser ablation of amorphous Teflon," Applied Optics, Vol. 32, No. 31, P. 6211-18, 1994.

68. Jervis, T.R., Nastasi, M., Hirvonen, J.P., "Excimer laser surface modification: process and properties," Beam Solid Interactions: Fundamentals and Applications Symposium, P. xvii+913, P. 665-78, 1994.

69. Hontzopoulos, E., Fotakis, C., Doulgeridis, M., "Excimer laser in art restoration," SPIE, Vol. 1810, P. 748-51, 1994.

70. Sunioka, K., Toyoda, K., "Modification and microfabrication of solid surfaces using excimer lasers," Riken Review, No. 1, P. 31-2, 1994

71. Matz, R., Heydel, R., Gopel, W., "In situ fabrication of InP-based optical waveguides by excimer laser projection," Applied Physics Letters, Vol., 63, No. 8, P. 1137-9, 1993.

72. Ebihara, K., Fujishima, T., Yamanata, Y., Ikegami, T., "Excimer laser ablation and plasma-enhanced MOCVD for YBaCuO superconducting thin film preparation," AIP Conference Proceedings, No. 273, P. 173-41, 1993.

73. Galindo, II., Vincent, A.B., Sanchez, R., Laude, L.D., "Excimer laser processing for surface improvement of tin oxide thin films," Journal of Applied Physics, Vol. 74, No. 1, P. 645-8, 1993.

74. Grebel, H., Fang, K.J., Manolis, C., "Paterned Schottyky barrier solar cells," SPIE, Vol. 1804, P. 126-9, 1993.

75. Alimpiev, S.S., Vartapetov, S.K., Vaselovskii, I.A., Likhanskii, S.V., Obidin, A.Z., "Shortening the pulses of KrF and ArF lasers during near surface breakdown in a liquid," Kvantovaya Elektronika, Moskva, Vol. 20, No. 3, P. 233-6, 1993.

76. Montagne, J.E., Sarnet, T., Prat, C., Inglesakis, G., Autric, M., "High-intensity KrF excimer laser processing of metal surfaces," Applied Surface Science, Vol. 69, No. 1-4, P. 108-14, 1993.

77. Niino, H., Yabe, A., "Excimer laser polymer ablation: formation into the metallization of polymer films," Applied Surface Science, Vol. 69, No. 1-4, P. 1-6, 2993.

78. Dye, R.C., Foltyn, S.R., Nogar, N.S., Wu, X.D., Peterson, E.J., Muenchausen, R.E., "Laser deposition and laser modification of high-temperature superconducting thin films," Proceedings of the Symposia on Reliability of Semiconductor Devices/Interconnections and Dielectric Breakdown for Microelectronic Applications, p. viii+330, P. 253-64, 1993.

79. Zach, K., Borck, J., Linzen, S., Krausslich, J., Schmidl, F., Schneidewind, H., Seidel, P., "Laser ablated YBCO thin films-relations between structural and electrical properties," Journal of Alloys and Compounds, Vol. 195, P. 199-202, 1993.

4.12. CROSS-SECTIONING DELICATE STRUCTURES

80. Dyer, T.E., Marshall, J.M., Pickin, W., Hepburn, A.R., Davies, J.F., "A comparison of polysilicon produced by excimer laser and furnace crystallisation of hydrogenated amorphous silicon (a-Si:H)," IEE Coloquium on 'Poly-SI Devices and Applications' Digest No. 67, P. 120, 17/1-4, 1993.

81. Feurer, T., Sauerbrey, R., Smayling, M.C., Story, B.J., "Ultraviolet-laser-induced permanent electrical conductivity in polyimide," Applied Physics A (Solids and Surfaces), Vol. A56, No. 3, P. 275-81, 1993.

82. Herman, P.R., Chen, B., Moore, D.J., Canaga, Retnam, M., "Photoablation studies of polymers, quartz and semiconductors with vacuum ultraviolet laser radiation," Photons and Low Energy Particles in Surface Processing Symposium, P. xv+552, P. 53-8, 1993.

83. Godbole, M.J., Pedraza, A.J., Lowndes, D.H., Thompson, J.R., Jr., "Excimer laser irradiation of metallic films on ceramic substrates," Phase Formation and Modification by Beam-Solid Interactions Symposium, P. xix+913, P. 583-8, 1993.

84. Smith, G.A., Chen, L.C., Chuang, M.C., "Effects of processing parameters on KrF excimer laser ablation deposited ZrO_2 films," Photons and Low Energy Particles in Surface Processing Symposium, P. xv+552, P. 429-35, 1993.

85. Okoshi, M., Toyoda, K., Murahara, M., "Excimer laser induced modifications of teflon surface into silicon carbide-like," Photons and Low Energy Particles in Surface Processing Symposium, P. xv+552, P. 377-82, 1993.

86. Allegrini, M., Arimondo, E., Callegari, C., Fuso, F., Iembo, A., Masciarelli, G., Berardi, V., Spinelli, N., Rossini, S., Danieli, R., "Laser ablation and deposition of graphite, lanthanum and lanthanum-doped fullerene," Proceedings of the First Italian Workshop on Fullernenes: Status and Perspectives, p., xi+233, P. 31-7, 1993.

87. Kononenko., T.V., Konov, V.I., Ralchenko, V.G., Strelnitsky, V.E., "KrF excimer laser etching of diamond-like carbon films," SPIE, Vol. 1759, P. 106-14, 1993.

Chapter 5

UV Optics and Coatings

5.1 Introduction

UV optical materials and coatings play a key role in laser imaging technology. The optical materials form the basis for manufacturing a wide variety of lenses, mirrors, and other critical optical elements needed to both generate and properly deliver UV energy to the desired surface. The coatings are similarly critical in maintaining the necessary transmission, reflection, and overall imaging quality in optical systems.

Coatings are used to protect a variety of refractive and reflective optical elements in beam delivery systems and in laser imaging lenses. They are also used on intracavity optics in excimer and other UV lasers (1). UV coatings are important in helping maintain high optical efficiency in laser beam delivery systems. Specially designed coatings are used to reflect and transmit specific wavelengths or wavelengths in a specified band.

Interference and antireflection coatings are widely used to increase peak transmission or reflection in a wide variety of optical imaging systems. They are also used on most all optical surfaces of laser imaging systems, including mirrors, attenuators, beam splitters, polarizers, windows, and other optics.

In this chapter we will discuss the various types of optical substrate materials and coatings used principally in excimer and ultraviolet laser beam delivery systems. We will discuss the key optical materials used as substrates for fabricating UV optical elements. We will also discuss the process for application of the coating, the specification and quality control aspects of UV coatings, and the functional properties of the coatings with various excimer laser wavelengths and energy levels.

5.2 UV Optical Requirements

Optical materials and coatings of special composition and specific design are required in order to provide optimal reflectance, antireflectance, and partial reflectance in ultraviolet imaging systems. The ultraviolet region of the spectrum poses special problems, namely: 1) high photon energies, 2) high degree of scattering, and 3) limited number of suitable substrate materials. The most common substrate materials onto which UV coatings are applied are listed in Table 5.1. These materials are used as lens elements for UV transmission.

	REFRACTIVE INDEX (AT 250nm)	COEFFICIENT OF THERMAL EXPANSION ($\times 10^{-6}$/°C)	CUTOFF WAVELENGTH (nm)
FUSED SILICA (SiO_2)	1.5076	0.5	147
CALCIUM FLUORIDE (CaF_2)	1.4676	18	123
LITHIUM FLUORIDE (LiF)	1.4188	32.3	105
BARIUM FLUORIDE (BaF_2)	1.519	18.4	134

Table 5.1: Deep-UV Optical Materials

The high photon energy at deep-UV wavelengths creates the need for very high purity lens materials and high quality coatings. Material defects in lenses or coatings become nucleation sites for absorption of UV photons, leading to heating effects, expansion, and fracturing of the material. High purity and high optical quality starting materials are therefore critical in producing damage-resistant optical systems for deep-UV laser imaging applications.

Deep-UV wavelengths are used because of the high image resolution possible and the ability to couple energy into substrates and substances to drive photoablation and photochemical reactions. For example, resolution < 0.15 μm is possible using a high numerical aperture lenses and imaging at 193 nm; using the same lens and optical parameters, a 436-nm (g-line mercury lamp) wavelength exposure would permit only 0.5 resolution. In the area of laser-assisted deposition and etching, UV laser systems permit a wide variety of applications not possible with longer wavelength lasers or other light sources (3). The increased interest in these areas has created the need for improved UV-coating technology and materials.

5.3 UV Optical Materials

UV optical substrate materials are fabricated into a variety of end products, including mirrors, attenuators, filters, beam splitters, and lenses. The basic substrate materials that can be used to produce these components are limited, as they must meet several critical specifications, especially high UV transmission and dimensional stability. UV substrate materials are used in transmission optical elements (attenuators, lenses, laser cavity optics/windows, beam-shaping optics) and reflection elements (beam splitters, laser cavity optics, mirrors). Some elements, like lens-mirrors, must meet the combined specifications of a lens and a mirror element.

In applications for transmission, the internal quality of the material is as critical as the surface finish, whereas in applications for reflection, only the surface quality is critical. In UV technology, maximum transmission of the elements is very important for total system optical efficiency. Lasers are expensive, and the manufacturing throughput factor in most applications is tied directly to optical efficiency. A plot of the transmission versus wavelength for typical UV substrate materials is shown in Figure 5.1.

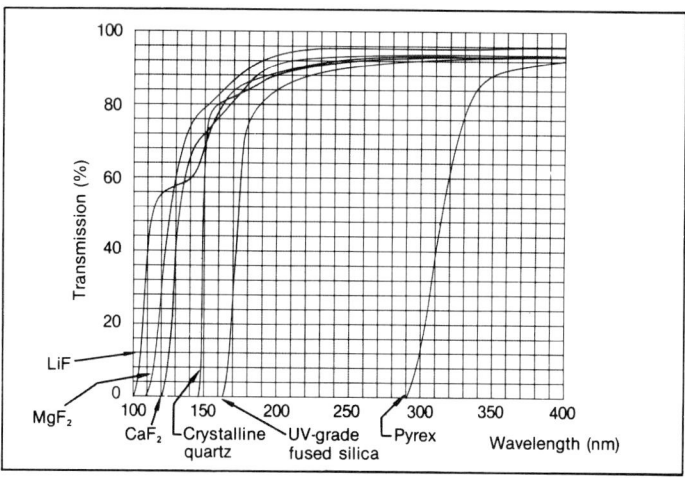

Figure 5.1: Transmission Properties of Deep-UV Substrate Materials

Note the lack of transmission of Pyrex below 300 nm, a factor restricting its use for front surface mirrors, where its dimensional stability and relative ease of polishing are desired properties.

5.3.1 Calcium Fluoride

Calcium fluoride (CaF_2) has excellent UV transmittance, surpassing that of fused silica. However, the lack of good thermal conductivity and fragility makes it undesirable for the fabrication of deep-UV lenses. Calcium fluoride will color center under intense deep UV radiation, forming first a blue color, and finally becoming opaque (5). Selecting the highest quality (maximum purity, transmission) calcium fluoride greatly reduces the tendency to solarize, but it is still mechanically fragile. Cerium-free CaF_2 is one grade that exhibits superior optical properties. Compared to magnesium fluoride (MgF_2), calcium fluoride is less costly and nonbirefringent.

5.3.2 Magnesium Fluoride

Magnesium fluoride is similar to calcium fluoride and transmits deep into the vacuum ultraviolet. Magnesium fluoride is more resistant to solarization than calcium fluoride, making it more useful for 193-nm excimer laser beam delivery or other laser optics. It will not discolor as easily as CaF_2, and if a tendency to discolor exists, it will show up in the first few thousand pulses.

5.3.3 Synthetic Crystalline Quartz

Synthetic crystalline quartz (SiO_2) transmits deep into the ultraviolet, reaching a limit around 180 nm (5). It has radiation damage characteristics similar to other glasses, but does not exhibit fluorescence or F-center discoloration. Synthetic quartz also becomes cloudy in the presence of fluorine when used in thicknesses suitable for excimer laser windows. In much thinner plates, for nonpressurized windows, it may be suitable in a fluorine ambient.

5.4 Radiation Damage

Some UV substrate materials are easily damaged by high energy UV radiation, such as lithium fluoride, which is highly hygroscopic and forms color centers (or solarizes) easily. Fused silica is especially resistant to discoloration caused by large doses of radiation. Electron bombardment causes a small (5%) decrease in the transmittance of fused silica, while gamma,

5.4. RADIATION DAMAGE

proton, and neutron radiation does not produce any measurable change in transmittance, as shown in Table 5.2 (6).

Radiation Type	Energy	Total Dose	Transmittance Test Results	Results
Electron	800 Kev	6.3×10^{16} Rads	$0.24 - 2.6\mu$	5% decrease in visible transmittance, no change above 2.0μ
	1.5 Mev	1.4×10^{7} Rads	$0.20 - 2.6\mu$	No change in transmittance above 0.3μ
Gamma	1 Mev	10^{8} Rads	$0.20 - 2.6\mu$	No change in transmittance above 0.45μ
		10^{10} Rads	$0.20 - 2.6\mu$	Slight blueish tint
Proton	2 Mev	10^{8} Rads	$0.20 - 2.6\mu$	No change in transmittance above 0.45μ
Neutron		10^{20} N/cm^{2}	$0.20 - 2.6\mu$	No change in transmittance above 0.35μ, 2% density increase

Table 5.2: Radiation Resistance of Fused Silica

The amount of laser damage is highly dependent on the UV laser or lamp source characteristics, including the beam profile, size, and pulse energy and pulse length (which influence thermal recovery time).

In applications requiring a high transmission at very short wavelengths, such as the windows at either end of an excimer laser, calcium fluoride is a logical choice, possessing both the durability and the optical transmission. The amount of internal defects, trace impurities, and contaminants also have an impact on the damage thresholds of UV materials. Laser-induced damage, in optics, especially in the UV spectrum, is caused primarily by impurities or material irregularities sometimes lodged in coatings that act as nucleating sites for UV absorption, leading to electronic excitation followed by expansion and explusion, part of the phenomenon of ablation.

Deep-UV optical materials have varying levels of hardness and will polish differently. The surface quality affects the performance of the optical coating. Imperfections in the surface can cause damage sites and also change the optical properties. Finished optics often have their edges ground to 45° to prevent chips of glass from redepositing on the surface where it could be encapsulated in a coating.

5.5 Coating Manufacture

Optical coatings are applied to a wide variety of optical substrates under high vacuum, typically $<10^{-6}$ Torr. Coatings consisting of dielectrics, metals, and semiconductors are used in UV laser applications. The dielectric coatings are composed of oxides, sulfides, tellurides, selenides, and fluorides (7).

UV optical coatings are deposited by either evaporation or sputtering. In thermal evaporation, a filament is heated until it vaporizes, after which it condenses onto the surface of the substrate in the vacuum chamber. Most UV coatings are applied by electron beam evaporation. A typical electron-beam evaporator schematic (26) is shown in Figure 5.2.

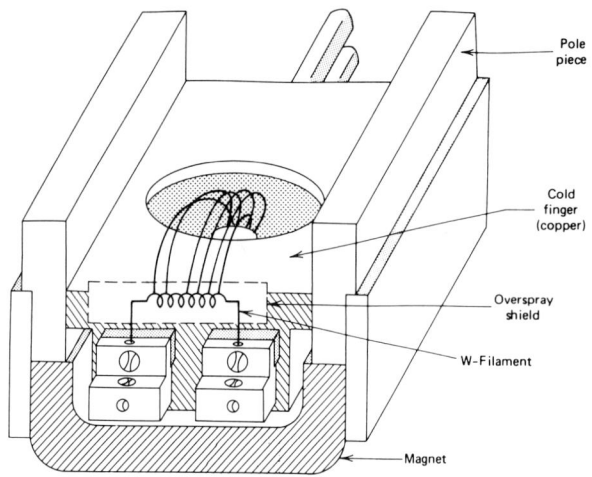

Figure 5.2: A Typical Electron Beam Evaporator Schematic

The material being deposited is resistance heated in an inert material container or boat usually composed of platinum, molybdenum, or tungsten. Dielectrics, which will vaporize at lower temperatures, may be heated directly or indirectly with a refractory filament. All of these deposition processes are classified as thermal "vaporization."

Vaporization by electron beam bombardment is another common coating method. Coating materials are heated directly by the high energy electron beam, directed into a small copper or similar inert crucible or multiple-pocket container if several different coating materials are to be deposited.

5.5. COATING MANUFACTURE

High temperature dielectric and refractory metals can be coated by this vaporization method.

The other main technique used to manufacture UV coatings is sputtering. In a plasma environment, the substrate is coated by the atoms of deposition material that are knocked off the "anode" by high energy heavy ions created in the plasma. The high energy associated with material deposition in sputtering helps to improve adhesion of the coating.

Coating thickness is measured during deposition by interference monitoring or resonance measurement. The interference method works by transmitting a collimated beam through a monitoring test plate mounted at the same plane as the substrate. The transmitted beam exits the vacuum chamber and is focused onto a monochrometer and photodetector. A lock-in amplifier chooses the chopped signal from the background; this signal is monitored for transmission maxima and minima. The original beam sent into the chamber comes typically from a continuous source, such as a tungsten halogen or deuterium lamp. The light from this lamp source is mechanically chopped and collimated.

The primary requirement for thickness measurement based on optical interference in thin film coating fabrication is a well-defined refractive index. Transmission or reflection of monochromatic light changes as the thickness of the deposited layer builds up. Dielectrics and semiconductor deposition materials typically have a refractive index higher than that of the substrate, a condition which will decrease transmission.

At coating thicknesses which occur at 1/4 multiples of the monitoring source wavelength, changes in transmission or reflection stop due to constructive (or destructive) interference points at maxima (or minima) intensity points. Minimum intensities occur at 1/4 wave multiples, while maximum intensities occur at 1/2 wave multiples. As the coating thickness increases, a proportionally smaller signal is either transmitted or reflected from the as-deposited structure.

Simple dielectric coatings (less than 4-5 layers) and metal coatings are thickness monitored by placing a quartz crystal plate in the same plane as the substrate during material deposition. Since the resonant frequency of the crystal is mass dependent, changes in mass during deposition can be used to determine coating thickness by monitoring resonant frequency changes. Too many layers on top of the crystal add variables that reduce the accuracy of correlation between the resonant frequency and thickness of the deposited layer, limiting this technique to just a few layers (7).

5.6 UV Coatings and Reflection

Thin film coating materials are applied to UV optics to increase or reduce light reflection. Electromagnetic theory shows that light reflection occurs when light moves from one medium (air) into another medium (glass) of higher refractive index (25). The amount of reflection is dependent on the polarization, angle of incidence, and refractive index. Increasing the refractive index causes increasing amounts of reflection. The reflectance for light passing from one medium (n_1) to another medium (n_2) of higher refractive index is given by the formula:

$$Reflectance = \frac{[(N_2/N_1) - 1]^2}{[(N_2/N_1) + 1]^2} \tag{5.1}$$

Reflection and light scattering is especially difficult in UV optical imaging systems because of the short wavelengths, ranging from about 150 nm up to 365 nm. Some applications in the extreme UV (XUV) involve wavelengths down to about 50 nm. UV light scattering and absorption are responsible for significant losses in efficiency in beam delivery systems, especially in the XUV.

The role of antireflection and interference or multilayer dielectric coatings is especially critical in maintaining maximum optical transmission throughout the system so that maximum energy is delivered to the substrate being irradiated. Significant improvements in optical transmission are possible by optimizing the reflection parameters. The two main ways to reduce reflection are to 1) match, as closely as possible, the refractive index of materials/mediums, and 2) coat the optical materials with a low refractive index film.

Light reflection and refraction on optical surfaces are governed by Snell's law (8), where the angle of incidence (i) and the angle of refraction (r) are specified as follows:

$$n_1 \sin i = n_2 \sin r \tag{5.2}$$

Figure 5.3 shows an example of light passing space through air (n_1) and onto an optical substrate (n_2) such as a lens element. Angle r shows the angle of refraction, angle "i" shows the angle of incidence, and the dashed line shows the path of the reflected ray.

n_1 in Snell's law represents the index of refraction of the area shown in the figure, and n_2 is the refractive index of the media into which the light is entering.

5.7. EXTREME UV COATINGS

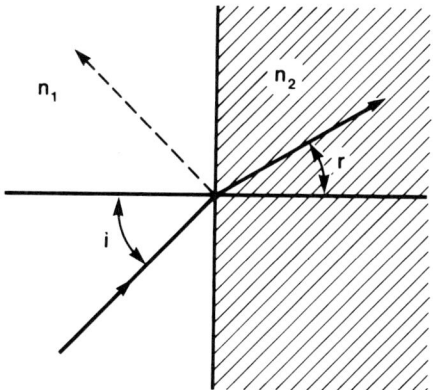

Figure 5.3: Light Reflection and Refraction on an Optical Surface

5.7 Extreme UV Coatings

The extreme ultraviolet (XUV) portion of the UV energy spectrum poses some unique problems. The XUV spectrum covers wavelengths in the 50- to 150-nm region. The high photon energies and scattering behavior of XUV wavelengths create special requirements for manufacturers of UV optics and coatings. Materials selected for XUV applications must exhibit high damage resistance, high internal (lenses) and external (mirrors, optics) purity, and be essentially free of occlusions or any defects (scratches, damage sites) that would serve as energy nucleation sites.

5.7.1 XUV Reflective Coatings

One type of reflective coating used on mirrors and UV optics is high reflectance iridium (27). The wavelength vs reflectance data in the plot shown in Figure 5.4 is for an XUV coating that was optimized for the 50- to 150-nm bandwidth. The solid line represents the iridium (a); the broken line shows reflectance for a UV-VUV coating of magnesium fluoride over aluminum (b). Iridium is generally recommended for reflectance with wavelengths below 110 nm.

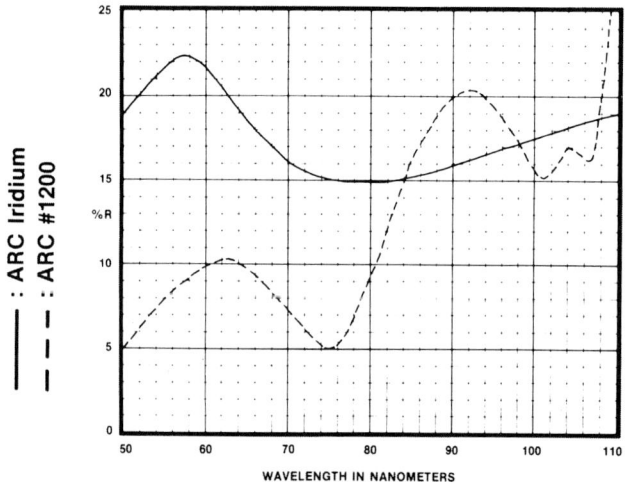

Figure 5.4: XUV Iridium Coating Reflectance Curve

5.7.2 XUV Beam Splitter

Beam splitters for the XUV operate by reflecting the XUV wavelengths while transmitting the longer UV wavelength radiation. A typical beam splitter is shown in Figure 5.5. In designing these optics, it is especially important to note the polarization states of the light; coatings are customized and optimized based on the polarization that will be used. Improper specifying of this parameter will result in loss of energy in the optical system.

5.7.3 Calculating Optical System Loss

Loss of energy in an optical system, especially UV laser systems, occurs when light is reflected away from the surface of (or absorbed in the bulk of) optical elements. The three main parameters for calculating the optical losses due to reflection are: 1) refractive index(s) of the material(s), 2) incident angle of the light, and 3) polarization of the incident light. The formulas are given as

$$R_p = \frac{\tan^2(i-r)}{\tan^2(i+r)} \qquad (5.3)$$

5.7. EXTREME UV COATINGS

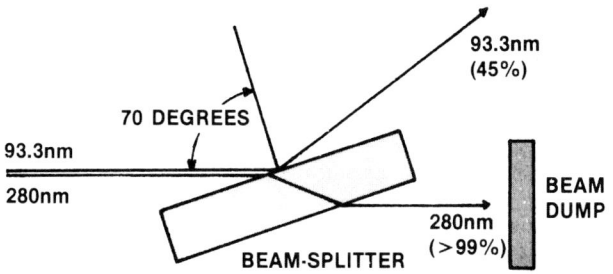

Figure 5.5: XUV Dichroic Beam Splitter

$$R_s = \frac{\sin^2(i-r)}{\sin^2(i+r)}, \tag{5.4}$$

where R_p = p polarized light
(or reflectance of light polarized
parallel to the plane of incidence)
 R_s = s polarized (or reflectance of light
polarized perpendicular to the plane of incidence

The above equations can be, at normal incidence, reduced to:

$$R = R_s = R_p = \left\{\frac{n_2 - n_1}{n_2 + n_1}\right\}^2 \tag{5.5}$$

In a typical nitrogen-purged UV imaging system, we could assume a refractive index of $n_2 = 1.5$ and $n_1 = 1.0$ for the ambient. Typical loss of light from reflectance is 4% per surface, or approximately 0.08% per element (7).

Taking an example of a lens system with 10 elements, the calculation for reflection loss (R_L) is:

$$R_L = 1 - (.96)^{20} \text{ or about 50\%}$$

Losses from reflection will increase as the refractive index of the material (n_2) increases relative to air (n_1). Also, in cases where $n_1 \geq n_2$, the reflected

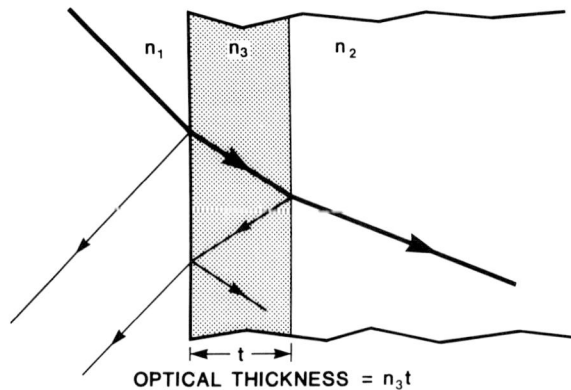

Figure 5.6: Optical Coating/Substrate Reflection Ray Paths

wave will undergo a phase change of 80°. Finally, ghost images may also be generated in uncoated optical systems. Antireflection coatings typically eliminate ghost images and most of the reflected light.

5.8 Single Layer Antireflection Coatings

Antireflection coatings are used to improve optical transmission efficiency. In this section, we will discuss the use of a single layer coating and its effect on optical performance in a system. Figure 5.6 (7) shows a schematic of an optical coating on a substrate, indicating the direction of light rays into and out of an optical element.

In the figure, n_1 represents an air or nitrogen ambient; n_2 represents the substrate onto which the dielectric coating is deposited; and n_3 represents a single layer antireflection coating. The first reflection occurs at the air/coating interface and the second reflection occurs at the coating/substrate interface; light from the second reflection bounces off the back surface of the dielectric coating, sending light back through the coating where, with sufficient intensity, it could produce a series of identical reflections within the coating.

5.8. SINGLE LAYER ANTIREFLECTION COATINGS

One technique to reduce reflection is to deposit the coating to the exact 1/4 wave optical thickness, thereby creating destructive interference in the reflected rays. The second ray, reflecting from the coating/substrate interface, will be 180° out of phase with the first reflected ray. If the thickness of the deposited coating is increased to a 1/2 wave thickness, the second reflected ray will undergo a 360° phase change and will therefore match the reflectance of the original, uncoated substrate.

Reflectance reduction in optical systems is treated mainly through the use of coatings applied directly to the optical elements. Antireflection coatings vary widely in composition, method of application, and useful wavelength range. In this section we will discuss single and multilayer optical coatings that are used mainly in UV optical imaging systems. Specific coatings are used for infrared, visible, and UV portions of the electromagnetic spectrum, each with specific optical properties customized for that spectrum.

Since antireflection coatings are based on 1/4 wave multiples, the amount of reflection needs to be calculated for every wavelength, and a coating optimized to satisfy the conditions of the optical system. The typical values for reflectance at various UV wavelengths are shown in Table 5.3. These are based on normal incidence, using 1/4 wave coating thickness. Reflectance reduction can be achieved over very narrow or wide spectral ranges.

In addition to wavelength dependence, reflection is both polarization and angle-of-incidence dependent. For example, most calculations are for normal incidence, but off-angle incidence will require more detailed and complex calculations. Optical system designs call out many different angles of up to 90°, but many designs have incident angles below 25°. At 25° angles, there is a shift in the reflectance curve toward shorter wavelengths, but little major change in the amount of reflectance. Changes in polarization, especially from unpolarized to S or P polarization, will impact reflectance.

5.8.1 Reflectance Calculation

The intensity maxima and minima reflecting from an optical surface are a function of the optical thickness, determined by multiplying the refractive index by the thickness of the medium or optical element.

In order to calculate the reflectance from any 1/4 wave thick film on a nonabsorbing substrate, the formula is given as follows:

$$R = \frac{(n_1 n_2 - n_3^2)^2}{(n_1 n_2 + n_3^2)^2} \tag{5.6}$$

Laser Wavelength (nm)	Reflectance at Normal Incidence (%)
193 (ArF)	0.75 to 1.5
212 (ND:YAG)	0.75 to 1.25
222 (KrCl)	0.5 to 1.0
248 (KrF)	0.2 to 0.5
266 (ND:YAG)	0.2 to 0.5
282 (XeBr)	0.2 to 0.5
308 (XeCl)	0.2 to 0.5
325 (He-Cd)	0.2 to 0.5
337 (N2)	0.2 to 0.5
351-353 (XeF)	0.2 to 0.5
355 (ND:YAG)	0.2 to 0.5

Table 5.3: Wavelength versus Reflection Using an AR Coating (Courtesy of Acton Research Corporation)

When $n_2 n_1 = n_3^2$, reflectance is zero. In terms of matching a refractive index value to a lens or optical element, the material in question would be $n_3 = 1.22$. This assumes $n_1 = 1.0$ (air), and the glass substrate is $n_2 = 1.5$. A typical value for BK7 glass is 1.5167 and for magnesium fluoride is 1.38%. The reflectance value for these materials is 0.0128%.

Reducing reflection can also be done by applying magnesium fluoride on flint glass (index $n_2 = 1.7$). This combination reduces reflection from 0.067% (uncoated) to 0.003% (coated) (7).

5.9 Multilayer Antireflection Coatings

Dielectric stacks or multiple-layer antireflection coatings are used for maximum reflection reduction and are more costly and complex to produce. Multilayer dielectric coatings are the primary technique used in UV laser optical systems for improving transmission and image quality, mainly because the short wavelengths of the deep-UV spectrum scatter so easily.

Multilayer antireflection coatings are functionally designed to perform the same role as single layer coatings: reduce the energy losses at the op-

5.9. MULTILAYER ANTIREFLECTION COATINGS

tical surfaces to near-zero. Multilayer coatings are used in place of single layer antireflection coatings. They provide improved transmission and wider wavelength ranges.

The primary design goals for multilayer, antireflection coatings take into account the many functional requirements of the optical system. Coatings need to be inert to optical system ambient, which may include nitrogen, argon, or other purge gasses. Ideally, the coating would function to its specifications regardless of the polarization state, wavelength, and angle of incidence. In practice, this is seldom the case, and coatings need to be customized for some or all of these parameters.

The energy density tolerance is important, especially in high power UV laser systems, and coatings are also customized or specially treated to withstand varying amounts of power density. The main design goal is antireflection, and typically other parameters are traded in order to optimize this single goal. For example, certain coatings will improve antireflection properties, but reduce power density tolerance.

Several multilayer techniques are used to achieve high degrees of antireflection, including multiple layers of varying thickness (with a single composition coating material) or multiple layers that have identical thickness but varying in composition. Combinations of similar and varying compositions are also used in layers whose thickness must be computer optimized due to the complexity of light interactions.

One type of multilayer, antireflection coating utilizes two layers of the same composition, with varying thickness, separated by a middle "absentee" layer. This combination is shown in Figure 5.7 (7,9).

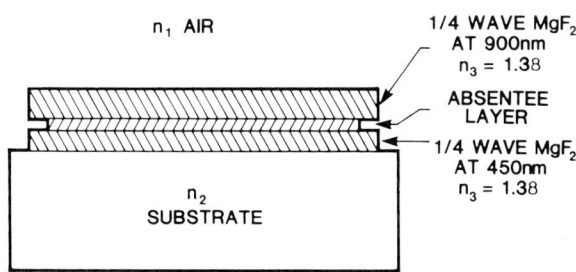

Figure 5.7: Multilayer Antireflection Coating

The first layer in Figure 5.7 is a quarter-wave-thick magnesium fluoride layer at 450 nm. The second "absentee" layer is used to return the index of refraction back to that of the underlying substrate. The top or third layer is also quarter-wave-thick magnesium fluoride at 900 nm and is half-wave thick at 450nnm. This coating will have two minimum reflection wavelengths, one at 45 nm and another at 900 nm. It will also provide reduced reflection (below that of a single antireflection layer) at the wavelengths between 450 and 900 nm.

A simpler example of a multilayer, antireflection coating is the "double quarter" coating, shown in Figure 5.8. The reflectance from this type of system at normal incidence is given by the formula (7):

$$R_{mn} = \frac{((n_3)^2 - n_1 - n_2(n_4)^2)^2}{((n_3)^2 n_1 + n_2(n_4)^2)^2} \tag{5.7}$$

This has a minimum at $n_4 = n_3(n_1/n_2)^{1/2}$, since n_1 is usually approximately 1, $n_3 > n_4$ to satisfy (8).

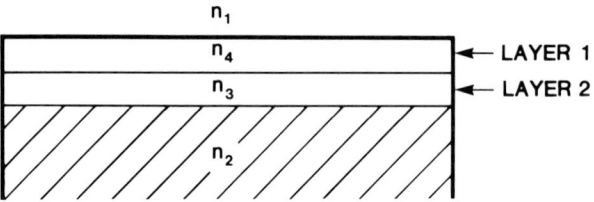

Figure 5.8: "Double-Quarter" Antireflection Coating

The glass substrate is coated first with a high index material, such as zirconium dioxide (n=2.1), zinc sulfide (n=2.32), or titanium dioxide (n=2.4). The second layer is a low index material, such as magnesium fluoride (n=1.38) or cerium fluoride (n=1.63). This type of coating is relatively easy to apply and is less costly than the three-layer structure shown earlier.

5.10 Dielectric Reflector Coatings

The use of multilayer dielectric coatings permits the reflectance of a variety of wavelengths as high as 99.9% by stacking a number of alternating layers of high and low index material layers. The first dielectric layer has

5.10. DIELECTRIC REFLECTOR COATINGS

a higher refractive index than the substrate (2.2 in the example) which is n=1.5. This raises the reflectance value of the quarter-wave layer to 0.314%, a considerable increase over the 0.042% of the glass substrate.

Figure 5.9 shows a 13 layer dielectric stack coating with 99.9% reflectance at the design wavelength (7). The second layer above the glass is a low index material (n=1.35); this and all subsequent layers are a quarter wave-thick. The second layer acts as an antireflection layer for the first layer, producing reduced reflectance. Reflectance is still much higher than the glass, but lower than the value achieved when the first layer was added. The third layer has an index higher than the second layer and acts as a reflector to the second layer, thereby increasing the total reflectance. This type of coating is very wavelength dependent, and its reflectance at other than the designed wavelength will be reduced (10).

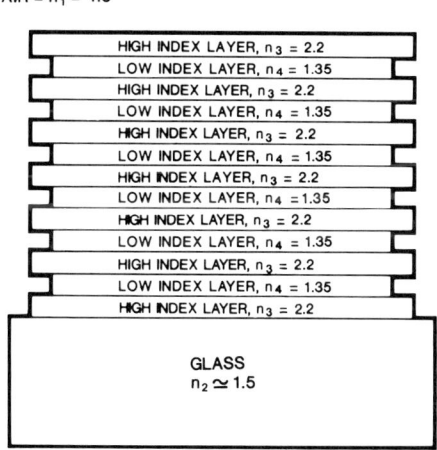

Figure 5.9: Multiple Layer Dielectric Coating

Dielectric reflector coatings have multiple uses in optical systems. Applications include attenuators, partial reflectors, beam splitters, mirrors (cold and hot), and short and long pass filters.

The curves show the transmittance at both 0 and 45 degree incidence. The range of transmittance (or reflectance) can be increased by using two separate dielectric stacks together, each stack representing a specific wavelength range. The coating stacks can be optimized to block all but very specific wavelengths, or transmit all but a small portion of the spectrum, depending on the application. Blocking the reflectance, for example, is accomplished by adding layers that are odd, quarter-wave multiples to the coating stack.

5.11 Metal Reflector Coatings

Metallic reflector coatings are thin, vacuum-deposited metal films that are applied to highly polished substrates. The reflectance values achieved with these films are very high, improved by the high degree of polish produced on the glass substrate. Metallic reflector coatings are relatively inexpensive and cover a wide spectral range (7). The types of metals used for metallic reflector coatings are shown along with their reflectance as a function of wavelength in Figure 5.10.

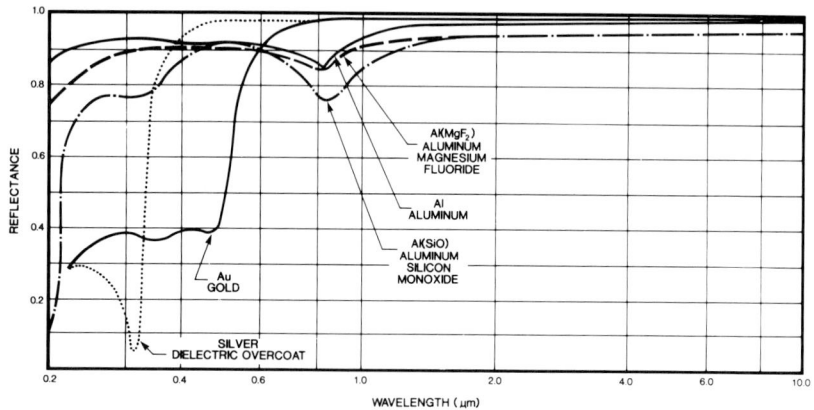

Figure 5.10: Reflectance of Metallic Reflector Coatings

The main metals used for optical mirrors are aluminum, chrome, silver, and gold, as shown in Figure 5.10. The loss of reflectance that occurs is due primarily to absorption of energy by the coating itself. Prior to deposition, the glass substrates are highly polished and checked for wave flatness. Substrates are cleaned just before deposition, as airborne contamination can

5.11. METAL REFLECTOR COATINGS

become encapsulated and impair coating quality.

5.11.1 First and Second Surface Reflectors

Coatings can be applied to either the face of the optic (first surface reflector) or to an internal surface of the optic (second surface reflector). The use of a second surface coating allows protection of the coating from scratches and abrasions that can occur during cleaning of the optics. Metallic coatings used on the front surface of substrates will react with ambient oxygen and tarnish or oxidize, especially aluminum and silver. This is prevented by overcoating the metal with a thin layer of optically compatible material (7). Figure 5.11 shows the principle of first and second surface reflector coatings.

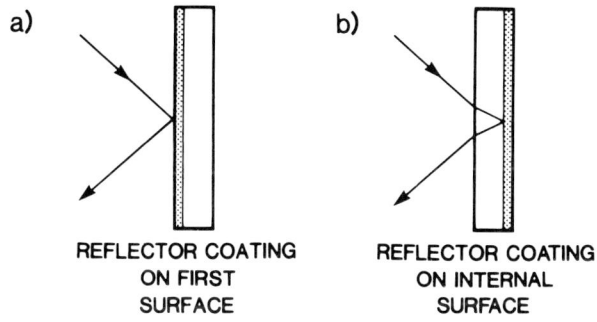

Figure 5.11: First and Second Surface Reflector Coatings

5.11.2 Aluminum Reflector Coatings

Aluminum mirrors are widely used in laser optics. They are relatively inexpensive to produce and simple to manufacture. Aluminum is less complex than the use of multilayer dielectric stacks, and overcoating of aluminum provides the needed protection against laser damage. Aluminized mirrors are used for beam delivery (beam turning and steering), final lens imaging (Schwarzchild objectives), and intercavity optics in excimer lasers.

Aluminum mirrors for excimer laser applications are required to withstand high laser beam fluence in the wavelength range of 351 to 193 nm. Intercavity laser mirrors must be able to withstand the corrosive gas mixtures (argon fluoride, krypton fluoride, xenon chloride, etc.) used to generate

Parameter	Value
Purity	>9.99%
Deposition Rate	300-100 Å/sec
Coating Thickness	500-2000 Å
Substrate Temperature	<100°C
Ambient Pressure	$10^{-5} - 10^{-6}$ Hz

Table 5.4: Aluminum Film Deposition

UV photons in the discharge chambers. Mirrors inside and outside laser cavities need to maintain high reflectivity so that UV photon losses are kept to a minimum.

High reflectivity aluminum mirrors are made by first fabricating a highly polished surface. Highly stable materials, such as "zero" expansion glasses, are preferred for critical applications involving high resolution imaging or beam positioning. The polished glass or quartz is cleaned, alcohol rinsed, and baked to remove surface moisture just before being placed in the aluminum evaporator. Ideally, the moisture bake-out can be performed inside the vacuum environment of the evaporator (14). Conditions for high reflectivity evaporated aluminum films are shown below.

High purity of the aluminum is important in achieving high reflectivity. Impurities, such as iron or copper, are introduced during evaporation or sputtering, and as little as 1% or less of these impurities reduce reflectivity 20-30% (15).

The deposition rate also affects reflectivity, where low rates will result in a loss of reflectivity. The same is true for ambient pressure, as higher pressures than those indicated above reduce reflectivity. Typical aluminum deposition parameters and corresponding values are shown in Table 5.4.

The substrate temperature plays a key role in the efficient bonding of the aluminum to the substrate. Temperatures too high may produce other species (hydroxyl groups) between the aluminum and the glass substrate, but a certain minimum temperature is required for good metal-glass interfacial adhesion. Some of this is related to the elimination of surface moisture (16).

The natural coating for evaporated aluminum is aluminum oxide, a "na-

5.11. METAL REFLECTOR COATINGS

tive" film that grows on the virgin aluminum surface as it scavenges oxygen from the air. Aluminum oxide (Al_2O_3) has a refractive index of 1.7 and only slightly reduces (0.1 - 0.2%) reflectivity (17).

The theoretical reflectance of a 100 Å thick layer of aluminum is 93.2% at 248 nm, using a 1/2 wave dielectric overcoat with a refractive index of 1.5. The reflectivity of the aluminum can be increased by the addition of a second dielectric layer. The two layer, quarter-wave stack (with a low of 1.45 and a high of 2.0) results in a calculated reflectance of 96.3%. Dielectric stacks, with many individual layers (up to 20 or more), are used in applications where extremely high damage thresholds (several Joules/cm^2) are required (18-20).

5.11.3 Thorium Fluoride Coatings

Thorium fluoride dielectric coatings are used as protective overcoats on aluminum mirrors for deep UV beam delivery, especially at 248 nm. Thorium fluoride (ThF_4) and magnesium fluoride (MgF_2) coatings are well suited for use as inner-cavity optics in excimer lasers, providing greater chemical resistance to excimer gas mixtures than nonfluoride dielectric coatings.

Thorium fluoride coatings provide damage thresholds up to 0.5 J/cm^2 when used on aluminum films 2 μm thick; the coating thickness of the fluoride is typically 300 Å at 249 nm (11). Reflectivity of 86% was achieved under these conditions. A reduction of the aluminum thickness to 5000 Å reduces the reflectivity.

As with magnesium fluoride overcoats, the conditions used for deposition are important in producing special physical properties, such as hardness in the as-deposited film. When the laser beam begins to degrade the coating, discoloration and lifting or ablation of the surface will be observed.

5.11.4 Silicon Oxide Coatings

Silicon monoxide and silicon dioxide films, applied to aluminum mirrors by evaporation, are used to protect the mirror surface, adjust its refractive index, and improve both the reflectivity and ablation threshold. Since mirrors need to be cleaned periodically, a thin layer of SiO_2 is useful to provide cleaning resistance. Mirror cleaning is much the same as for other optical components: proponal alcohol or solvent swab, preceded by a mild optical detergent if soils are evident on the mirror surface. A final dry nitrogen gas cleaning is also recommended to remove airborne particles.

If scratches form, oxides can be stripped and redeposited, saving the expense of realuminizing. Silicon oxide overcoating, followed by UV irridation, has also produced increased reflectivity of aluminum (20), especially below 300 nm.

5.11.5 UV Conditioning of Mirrors

Simple UV sources, such as mercury discharge lamps, have been used to condition mirrors, providing improved reflectivity for the 248- and 193-nm excimer lines.

A commonly available source for UV for this purpose is a UV 141 EPROM eraser. Irradiation times of approximately 3 hr have been used to simultaneously reduce the refractive index and increase thickness of the oxide and the surface reflectivity (21).

An explanation of the physical changes in oxide films is that the SiO_2 reorders itself toward SiO_2 as oxygen is absorbed in the surface (oxidation). In addition, the film reorders itself somewhat during UV, and realignment of the crystal faces (polycrystalline structure) seems to favor higher reflectivity. In general, silicon-based oxide coatings are well suited for UV laser beam delivery applications.

5.11.6 Magnesium Fluoride Coatings

UV-reflecting mirrors are also overcoated with dielectrics other than silicon oxides to improve reflectivity and increase the damage threshold. Magnesium fluoride is suitable as a mirror overcoat, used as a half-wave dielectric layer for maximum reflection. Fluoride overcoats are preferred for intercavity optics in excimer lasers due to their resistance to chemical attack by argon fluoride and krypton fluoride. Magnesium fluoride can be deposited at room temperature, permitting easier removal of the MgF_2 in the event of laser damage. However, the coating must be dense and hard enough to provide the needed resistance to the UV beam.

Deposition conditions largely determine the characteristics of the as-applied film. A substrate temperature of 200-300°C is used to produce dense, hard films of MgF_2 (28). The higher temperature may affect the characteristics of the aluminum film, which has a specified deposition temperature of 100°; a compromise deposition temperature of 150°C is used.

A typical damage threshold for a room temperature-deposited film is about 0.5 - 0.6 J/cm^2, using a back-coated mirror tested through the quartz.

aluminum thickness was 1000 Å, with a MgF$_2$ thickness of 550 Å. UV laser damage threshold data for several materials are shown in Table 5.5.

MATERIAL	THICKNESS RANGE (mils)	THRESHOLD (J/cm^2)	APERTURE (μm)
Alsimag	0.10 - 10.00	9	50 × 50
Alumina	1.00 - 10.00	5	50 × 50
Aluminum	0.0005 - 1.00	6	50 × 50
Chrome	0.0005 - 1.00	1	50 × 50
Copper	1.00 - 5.00	9	50 × 50
GaAs	4.00 - 10.00	20	50 × 50
Germanium	4.00 - 7.00	20	50 × 50
Gold	0.0005 - 1.00	1	50 × 50
Inconel	0.0005 - 1.00	20	50 × 50
Indium Tin Oxide	0.0005 - 1.00	5	50 × 50
Lithium Niobate	0.0001 - 0.0002	3	50 × 50
Lithium Tantalate	0.0005 - 0.005	3	50 × 50
Nickel	0.0005 - 0.005	20	50 × 50
Niobium	0.0001 - 0.0002	3	50 × 50
Nitride	0.0005 - 0.005	5	50 × 50
Polycarbonate	1.00 - 2.00	15	50 × 50
Polyimide	0.001 - 10.00	0.03*	50 × 50
Polysilicon	0.001 - 0.100	3	50 × 50
Polysulfon	1.00 - 5.00	1	50 × 50
Sapphire	0.01 - 5.00	23	50 × 50
Silicon	1.00 - 20.00	20	50 × 50
Stainless Steel	0.001 - 4.00	12	50 × 50
Titanium	0.0005 - 0.002	12	50 × 50

Table 5.5: UV Laser Damage Threshold Data

Gold makes a good reflector coating from the near IR (800 nm) to the far IR (30 μm), with reflectance values above 98%. Reflectance exceeds 95% above 1.5 μm (7). Gold is essentially nonreactive with ambient air, even though a thin oxide will form over time. The main disadvantage of gold as an optical coating is its softness. Cleaning must be noncontact, using water, detergent, organic solvent, and clean, dry air.

5.12 UV Coating Types by Application

There are many applications for coatings used in UV optical systems, as well as a variety of materials used to make them. In this section we will describe specific optical applications, the optical substrates used, and the coatings for each application.

5.12.1 Attenuators

A UV beam delivery system may have 50 or more individual optical components and may have 10-12 different types of optical coatings. In a typical beam delivery system, optical attenuators are used to adjust the fluence levels in the optical path and to keep them below the damage threshold of the optics coatings.

Attenuators may be either fixed or variable. A fixed attenuator is a substrate coated with a multilayer dielectric (MLD) that is designed to reflect (or dump) away a given percentage of the beam energy, transmitting the rest through the system. Fixed attenuators can be added in series to increase the amount of attenuation.

Fixed attenuators are used as single optical elements and are manufactured to reflect a specific value (5, 10, 50%) and to transmit the balance of the energy. The drawback of using these elements is the number of undesirable surfaces they introduce into the optical path.

In some cases, as many as 3 or 4 elements are needed to achieve the required attenuation. Each optical surface will cause some loss of energy through reflection. A loss of 5% per surface results in losses of up to 60% for an optical path containing 14 elements (.95 X 14). A single variable attenuator is much more efficient, minimizing the energy loss. Figure 5.12 shows the principle of operation of the fixed and variable attenuator coatings.

Additional attenuator elements increase the difficulty of alignment and require additional cleaning. The thickness of the attenuator is also important. Increased thickness will cause undesirable translation of the beam. This can be eliminated by adding an opposing element at an opposite angle. Reducing translation of the beam is done by using thin optical elements. However, optics that are too _thin_ are more easily distorted or bent, and can cause a coating to reflect nonuniformly. A beam homogenizer can be positioned on either side of the attenuator(s).

Graded attenuators are optical elements which transmit a range of energy by changing the ansular position of the surface with respect to the beam.

A variable attenuator is a substrate of high UV transmission (high grade quartz, Supersil) overcoated with a MLD (multilayer dielectric) coating that changes its reflectance according to the angle of beam incidence. Variable attenuators help keep the number of optical elements to a minimum compared to graded or fixed attenuator elements.

The multilayer dielectric coating used as a variable attenuator can be designed to reflect a wide or a narrow range of wavelength or energy levels as

5.12. UV COATING TYPES BY APPLICATION

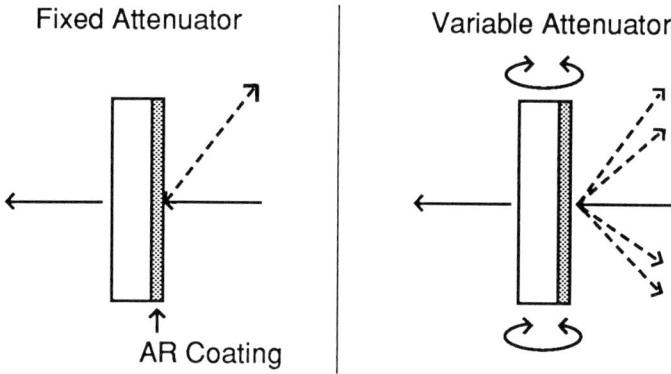

Figure 5.12: Principle of Fixed and Variable Attenuator Coatings

a function of beam incidence angle. Figure 5.13 shows a typical transmission plot of such a coating, taken at various angles of incidence.

A critical part of using such an element in an optical system is calibration and mechanical placement accuracy. Once the element is angle-calibrated, it is mounted in a holder. When moved to selected angles in the optical path, it properly reflects the energy value identified during calibration. This type of optical subsystem should be recalibrated periodically to verify performance at its critical angular positions. Extended use with high fluence may change the reflection characteristics of the coating. Regular cleaning of these optical elements will increase their useful life, as contaminants serve as nuclei for energy absorption. Also, the thickness of the element should be minimized without compromising dimensional stability or rigidity.

5.12.2 Homogenizers

Lens homogenizers are generally coated with an antireflection coating. One example is a tunnel homogenizer, a multifaced mirror device made up of multiple mirror elements, arranged in a square, triangular, hexagonal, or many complex shape, that bounce and mix or homogenize the laser beam energy. Coatings for mirror homogenizers are enhanced aluminum or alu-

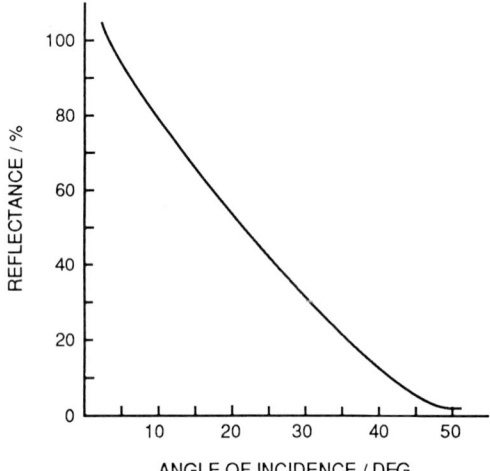

Figure 5.13: Reflectance vs Angle of Incidence of a Variable Attenuator

minum overcoated with magnesium fluoride.

The changes in reflectance of the different curves are due to the variation of the coating thickness of the magnesium fluoride. Aluminum with the magnesium fluoride overcoat is an excellent broadband reflector, ideal for mirrors in relatively low fluence applications. The typical absorption of this coating, on aluminum, is 10-15%. In medium fluence (5 - 20 J/cm^2) applications, the losses from absorption are undesirable and the likelihood of damage is greater. Multilayer dielectric coating stacks are used in these applications for maximum reflection.

5.12.3 Beam Benders and Beam Splitters

Beam turning and beam splitting are common functions for UV coatings. Beam turning requires maximum reflectance and may include transmission of another visible wavelength that will be used for visual observation or alignment. A multilayer dielectric beam-turning coating is made for high efficiency over a narrow spectral band, peaked or centered at the laser (or lamp) wavelength. The 193- and 248-nm wavelengths are the most common ones used in excimer laser industrial and research applications, and Figure 5.14 shows the reflectance versus wavelength plots for the 193-nm

5.12. UV COATING TYPES BY APPLICATION

wavelength.

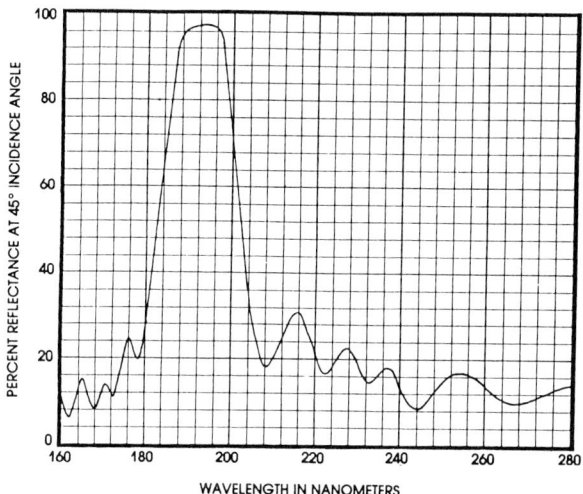

Figure 5.14: Reflectance vs Wavelength for a 193-nm Excimer Line

Coatings for beam splitters can be designed for multiple wavelength applications. For example, introducing a visible or operator viewing wavelength into an excimer laser delivery system will require two specific coatings on the beam splitter substrate. The visible coating will transmit approximately 50% of the light from the mercury, xenon, or other visible illumination viewing path; the excimer coating will reflect 98-99% of the laser beam energy. This permits division of the wavelengths to permit viewing coaxial with the laser illumination and avoids adding special optics for each path (29). It also keeps the sample in focus for both viewing and laser exposure, avoiding focus offsets.

The coating materials used to construct the dielectric layered films are tailored for a specific wavelength. Different materials may be used despite only several nanometers difference in their spectrum. Materials are selected that keep absorption to a minimum, as absorption reduces the life of the coating. Laser beam turning also requires maximum reflection.

5.12.4 Multiple Wavelength Coatings

Many UV optical systems require multiple wavelength transmission, where the UV laser wavelength is used to expose the sample. The second wavelength can be an alignment beam, such as a 633-nm helium neon laser, or a broadband mercury vapor lamp source used for visual observation.

The ideal two-wavelength coating will exhibit strong and specific reflection at the two chosen wavelengths and almost zero reflectance at all other wavelengths, as shown in Figure 5.15. Since the optical elements used for these applications are generally beam-turning mirrors, the coatings are optimized for these dual optical properties at a 45° angle of incidence.

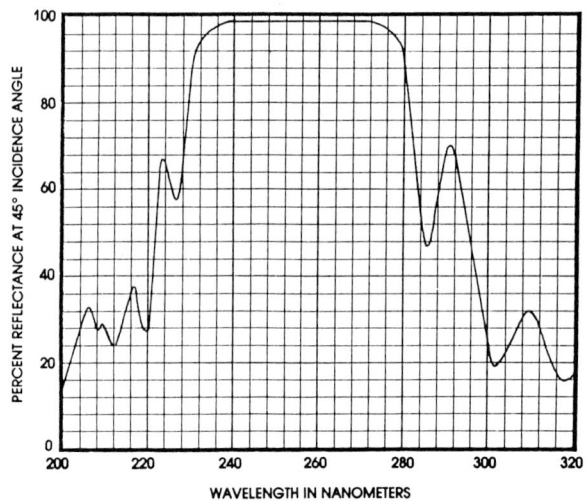

Figure 5.15: Reflectance vs Wavelength for a 248-nm Excimer Wavelength

Multilayer dielectric coatings make the most efficient dual wavelength mirrors; UV-enhanced aluminum will reflect very well in the UV (85-92%) as well as the visible, but has limited use in high fluence applications due to its relatively high absorption.

Multilayer dielectric coatings are first designed on computers, and after simulations are calibrated and verified functionally. Once optimized, MLD coatings are electron beam deposited onto laser-grade substrates for maximum reflection. The excimer wavelength will reflect above 90% in most cases, and the second wavelength (633-nm HeNe alignment laser, or simi-

5.13. COATING DAMAGE AND DEFECTS

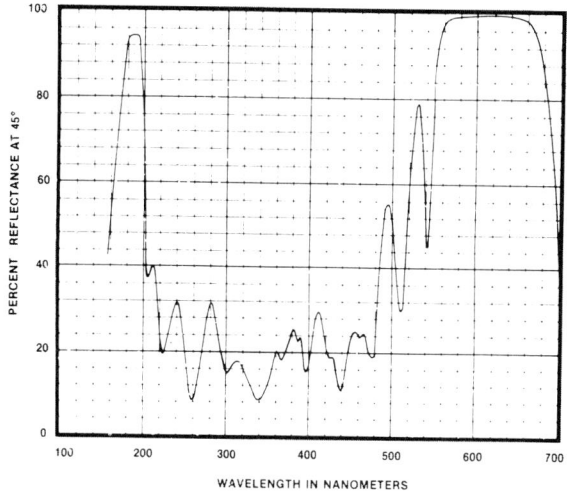

Figure 5.16: Dual Wavelength Beam-Turning Mirror

lar "visible" laser) will typically reflect at an equally high value (90-98%). The reflectance versus wavelength curve for a typical dual wavelength MLD coating is shown in Figure 5.16.

5.12.5 Dual Wavelength Beam Splitter

Another application for a coating in a deep UV laser imaging system is the transmission of a primary wavelength along with beam splitting of a visible wavelength to permit video camera recording during laser exposure. The optical properties of this type of element require maximum reflection of the laser wavelength on one surface of the element and medium to high transmission of some range of visible light through the back of the same element. In optical systems which require transmitting multiple wavelengths of light through the optical path, multiple wavelength coatings are required to obtain maximum light throughput or energy efficiency in the system.

5.13 Coating Damage and Defects

The high photon energies of UV wavelengths increase the criticality of coating quality and also increase the probability of coating damage. Three key aspects of this problem are: 1) the specific wavelength and its corresponding photon energy, 2) quality control in manufacture of UV coatings, and 3) sources of damage.

5.13.1 Wavelength Dependence

The key challenge for UV coating performance is the high photon energy and the fluence levels of excimer laser beams used in optical delivery systems. For example, the energy per photon at the 351-nm excimer line (XeF) is 3.53 eV, while the energy per photon at the 193-nm excimer line (ArF) is nearly double at 6.42 eV.

The high photon energies of ultraviolet wavelengths can cause compaction, deterioration, solarization, ablation, and eventual removal of UV coatings from optical surfaces. The 157-nm (fluorine), 193-nm (argon fluoride), and 248-nm (krypton fluoride) wavelengths are especially difficult, with damage thresholds as low as 1 J/cm^2. Much lower photon energies at the 308-nm (xenon chloride) and 351-nm (xenon fluoride) wavelengths permit energy densities on optical coatings as high as 20 J/cm^2 or more.

As the wavelength is reduced into the deep, vacuum (VUV), and extreme ultraviolet (XUV), cleanliness of the coatings becomes extremely critical. The incident energy nucleates around inclusions, voids, substrate defects, and other artifacts that impact the optical performance of a coating. As energy is intensely absorbed, heat is generated, causing expansion, cracking, and possibly ablation of the absorbing elements.

UV coatings must be tested with a range of energy levels at the wavelength in question in order to ascertain the damage threshold. The shorter UV wavelengths are also more critical because they are more intensely absorbed into the coating. High absorption around a particle in the coating causes a nucleation site or damage site. Once such a site is created it acts as a conduit for additional and increasing amounts of energy absorption, spreading the damage throughout the coating.

The origin of these sites can be a microscopic flaw (void or inclusion) in the coating surface. Figure 5.17 shows examples of these flaws detected with a scanning electron microscope at 10,000x magnification. The SEM photo was taken on the surface of a mirror lens taken from a Schwarzchild reflecting objective. The surface of this element is subjected to relatively high fluence excimer laser energy, and a high degree of coating quality is essential. The lifetime of these types of optics and coatings is typically specified in numbers of laser pulses.

5.13. COATING DAMAGE AND DEFECTS

Figure 5.17: Defect in a MLD Coating

5.13.2 Quality Control and Inspection

SEM analysis is a quality control technique that can be used prior to installation of optics into their housings. Figure 5.17 shows a typical type of flaw, an inclusion; voids are the other defects found (an inclusion), (A) and a void (B). Both types of defects will give rise to coating breakdown as energy from the laser tends to nucleate in these regions.

These flaws are only several microns across and less than 1000 Å deep, but because the optical surfaces on which they reside are intended for high fluence micromachining at 193 nm, they could easily initiate breakdown of the coating. Micron-sized irregularities and defects are difficult to find, especially at 10,000x magnification with a scanning electron microscope. Low magnification inspection can be performed first, either with an optical microscope or with the SEM at 200x.

Scattering of the short ultraviolet light wavelengths also poses special challenges for UV optics designers and manufacturers. A given optical element (lens, beam splitter) may lose up to 5% per surface without an antireflection coating. This can be reduced to 1-2% with an antireflection coating. Scattering and ghost images also detract from the resolution and image fidelity in these systems.

5.13.3 Sources of Damage

Optical coatings used on lenses, mirrors, beam splitters, and other UV optical imaging elements are likely to undergo degradation with continued use. Optical damage is also caused by other factors that singularly or collectively lower the optical damage threshold. These include mechanical abrasion from cleaning, photobleaching, coating nonuniformities, inclusions, voids, chemical corrosion, oxidation, and thermal effects.

Abrasion can be caused by cleaning optical elements, where small microscratches are left on the coating surface from mechanical abrasion. If a small particle contaminant is also located on or near a scratch, the damage threshold may be lower due to the coincidence of two defects.

Nonuniformities in the coating may also cause degradation of a coating. Changes in coating thickness give rise to changes in the thermal conductivity of the film. Heating effects not only accelerate optical damage, but induce other negative effects, such as photobleaching. Photobleaching may give rise to additional heating due to increased optical transmission of the incident photons.

Many UV optical coatings have very complex optical and chemical structure. Variations in the stoichiometry of the film may result in reduced damage thresholds or other nonlinear effects.

Some specific examples of UV laser damage include ablation of holes in the coating, complete removal of the top layers of the film, severe discoloration, cracking, crazing, and compaction of the surface. All of these effects will degrade the optical performance of the coating, but may or may not be considered critical, depending on the application. A working definition of optical damage states that "The damage causes an unacceptable functional change in the performance of the optical element or system in question."

5.13.4 Damage Threshold Testing

There are several methods for measuring UV laser-induced optical damage on surfaces. Early techniques relied on averaging the number of nondamaging and damaging fluence levels for a given set of conditions. These tests worked well for a given beam diameter or focal spot size, but were useless in predicting conditions for any other size or for any change in test parameters.

The more commonly used method for damage threshold testing is to plot the percentage of sites that undergo optical damage versus the beam fluence. The damage threshold is then defined as the 0% crossing of a curve fit to

5.13. COATING DAMAGE AND DEFECTS

the data. An example of the data from this test method is shown in Figure 5.18. The data in the figure are derived from testing of optical elements after optical coatings are applied to the element and specifications verified by a test piece inserted into the vacuum chamber with the optics during a coating run (15).

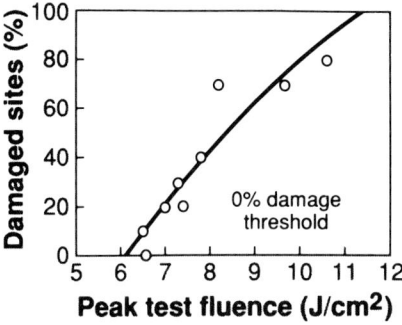

Figure 5.18: Damage Threshold Plot

In any test of damage threshold of a coating used in a UV optical imaging system, it is critical that all of the test parameters be precisely given, as slight variations in any of the key parameters will result in changes in the damage threshold figures. For example, very slight changes in thickness of an aluminum film, lying under a UV coating being tested, will change the thermal conductivity values and hence the absorption properties of the coating on top of the aluminum.

When the optical thickness (a physical distance change that causes a change in the wavelength of light at the test point) changes, especially for a quarter multiple of the wavelength, the intensity value of incident light will change, and possibly change the damage threshold value.

5.13.5 Laser-Induced Damage Mechanisms

There are several distinct steps in the interaction mechanisms that lead to coating damage from laser beams. The first step is absorption, occurring when the incident UV laser energy is deposited first on the surface of the film, and secondly within the bulk of the film. Absorption rates are driven mainly by the absorption coefficient of the material and by the wavelength of the incident UV energy.

As a general rule, the higher the absorption coefficient of the material, the lower the damage threshold since absorption gives rise to the second mechanism at work in creating damage: heat generation and conduction away from the damage site.

Typically, damage sites occur as isolated areas on the substrate material and not as a uniformly distributed phenomenon. Continued exposure of the damage site to the laser beam results in a buildup of heat at a rate greater than the rate of conduction of heat away from the site.

These three key mechanisms, absorption, conduction, and heating, are interdependent and nonlinear. For example, the rate of absorption of energy in a sample will change as a function of the amount of heat generated over time. Also, the rate of conduction of heat away from the damage site will change rapidly when excessive heating leads to ablative decomposition, completely removing a portion of the film.

Advances in coating technology have led to the development of optical coatings that will withstand several Joules/cm^2 of deep-UV laser fluence on the surface of the optical element. Studies on the mechanisms of laser-induced damage to both optical coatings and optical elements have also led to a better definition of the actual chemistry and process of laser degradation of optical surfaces.

5.14 Coating versus Polarization

In many laser beam delivery systems, the beam needs to be reflected in both a horizontal and in a vertical plane. Beam-turning mirrors and beam splitters that allow the introduction of separate optical paths are used to direct the laser energy. They are typically coated according to the wavelength used. The coating needs to be matched to the polarization of the laser beam.

Polarization essentially describes the alignment of the magnetic and electric fields in the laser beam. While polarization can be of several different types or orientations, the common types for laser optics are S-polarization and P-polarization. These types of polarization define how the light wave strikes an optical surface. S-polarization means that the polarization is parallel to the SURFACE of the mirror that reflects the light. P-polarization means that the polarization is in the PLANE of the incident and reflected rays.

Lasers with horizontal polarization will therefore use all P-polarization mirrors if the beam remains horizontal and S-polarization mirrors when the

5.14. COATING VERSUS POLARIZATION

beam is reflected up or down after it exits from the laser. Figure 5.19 illustrates the S-polarization and P-polarization principles.

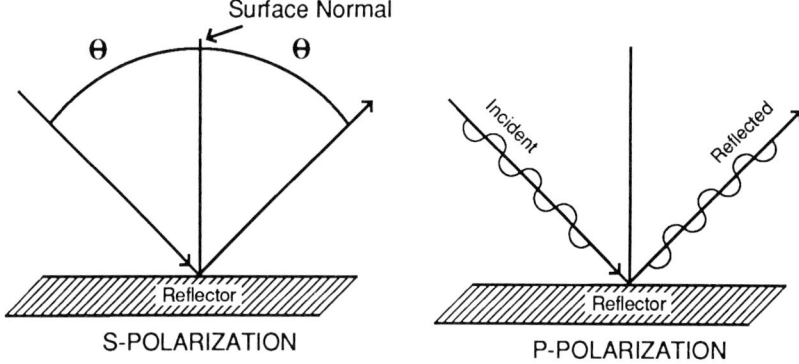

Figure 5.19: S-Type and P-Type Polarization

Coating specifications are especially critical in beam delivery systems in order to conserve beam energy, reduce the amount of scattered light, and properly filter the desired wavelengths. Polarization is one of many specifications used to manufacture beam delivery and laser cavity optics.

5.14.1 Coating Damage versus Repetition Rate

Coating damage in laser optical delivery systems is strongly dependent on laser repetition rate. As a rule, an increased repetition rate results in lower damage thresholds as the coating has less time to recover electronically and thermally. Increased amounts of energy deposited in a film result in a higher level excitation state. If the heat is not dissipated rapidly enough, it causes expansion, cracking, and ablation of the film when the damage threshold is exceeded.

In many organic coatings, the initial pulse or two pulses will bleach, compact, and may chemically rearrange the surface of the film. These effects have been observed in photoresist films, causing changes in refractive index as a function of dose. Figure 5.20 shows the refractive index of a positive-working, novolak-resin based photopolymer before and after exposure to a near UV-mercury vapor source (365 nm).

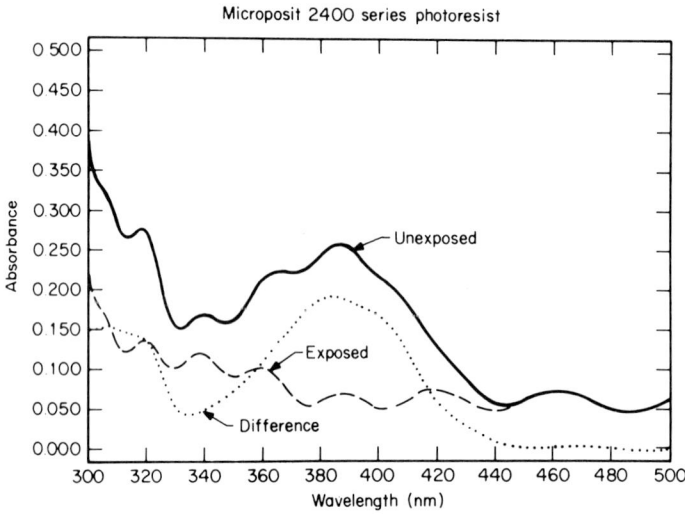

Figure 5.20: Refractive Index of Photoresist (Exposed and Unexposed)

Following exposure to a single or two laser pulses, many materials exhibit increased damage thresholds since the initial pulse(s) at low rep-rates (1-10 Hz) provided a conditioning effect in the film. The initial pulses may also be responsible for removing surface contamination or oxidation layers. Characterization of the damage threshold of a given material must account for the changes that occur in the initial pulses. As the rep-rate is increased gradually, many materials are able to withstand increased energy doses/unit time due to the conditioning effects of initial pulses.

5.15 Optical Fiber Beam Delivery

Optical fiber beam delivery is accomplished with fused silica and, in some cases, polymeric materials. Launching high intensity UV laser pulses into these fibers is complicated by the problems of front and rear surface damage thresholds and "on-center" collection of the light at the input end. A typical damage threshold is about 20 J/cm^2 at 308 nm, using a 25-nsec pulse (30). The schematic of a tapered fiber in Figure 5.21 helps solve both collection and surface damage problems by offering a funnel to input the light which simultaneously spreads the energy to prevent ablation. This concept was

5.15. OPTICAL FIBER BEAM DELIVERY

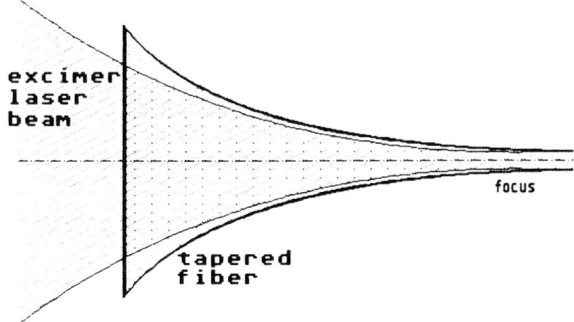

Figure 5.21: Tapered Fiber for Higher Pulse Energy Delivery (Courtesy of Lambda Physik)

developed by K.O. Greulich and H. Hitller at the University of Heidelberg.

References

1. Elliott, D., Pennelli, C., Sengupta, U., "Excimer Laser for Microlithography," 3-Beams Conference, Seattle, 1991. J. Vac. Sci. Tech., 1992.

2. Markle, D.A., "Deep UV Lithography: Problems and Potential," SPIE Vol. 774, p. 108, 1987.

3. Ehrlich, D., "Early applications of laser direct patterning: direct writing and excimer projection," Solid State Technology, Dec. 1986.

4. Case, M., Flint, B., Merk, R., Richards, K., "Mirrors for Excmier Lasers," Lasers and Applications, March, 1985, p. 85-89.

5. "Excimer Optics: Short Wavelengths Require Careful Material Selection," Laser Focus Workd, March, 1983, p. 20-22.

6. Ackerman, B., Corning Glass Works, Corning, New York; Technical Reference 607-974-4240, September, 1988.

7. Dr. Eugene G. Arthurs, Thom Connolly, Oriel Corporation, Stratford, Conn.; Technical Publication: Optics and Filters, 1990.

8. Hecht, J., Ed., "Understanding Lasers," IEE Press, Piscataway, NJ, 1991.

9. Callahan, G., Acton Research, Private Communication.

10. Atwood, B., Uvira Optics, Private Communication.

11. Elliott, D.J., Piwczyk, B.P., "Excimer Lasers in Micromachining," Proceedings, Materials Research Society Meeting, Boston, 1987.

12. Case, M., Flint, B., Merk, R., Richards, K., "Mirrors for Excimer Lasers," Lasers and Applications, March, 1985, p. 85-89.

13. Case, M., "Optical Coatings for the Deep UV," The Photonics Design and Applications Handbook, 1989.

14. Lewis, K., "Ultrahigh Vacuum Processing Yields Ultrahigh Quality Coatings," Laser Focus World, p. 127, January, 1990.

15. Hass, G., Junter, W.R., Tousey, R., "Reflectance of Evaporated Aluminum in the Vacuum Ultraviolet," J. Opt. Soc. Am., Vol. 46, No. 12, Dec. 1956.

16. Hass, G., et al., "Influence of Purity, Substrate Temperature and Ageing Conditions on the Extreme Ultraviolet Reflectance of Evaporated Aluminum," J. Opt. Soc. Am., Vol. 47, No. 12, Dec. 1957.

17. Madden, R.P., Canfield, L.R., Hass, G., "On the Vacuum Ultraviolet Reflectance of Evaporated Aluminum before and during Oxidation," J. Opt. Soc. Am., Vol. 53, No. 5, May, 1963.

18. Hass, G., "Filmed Surfaces for Reflecting Optics," J. Opt. Soc. Am., Vol. 45, No. 11, Nov., 1955.

19. Behrens, H., Ebel, G., eds., Physics Data, Optical Properties of Metals, Part II, Badendruck, 1981.

20. Bradford, A.P., Hass, G., "Increasing the Far Ultraviolet Reflectance of Silicon Oxide Protected Aluminum Mirrors by Ultra-violet Irradiation," J. Opt. Soc. Am., Vol. 53, No. 9, Sept. 1963.

21. Bradford, A.P., Hass, G., McFarland, M., Ritter, E., "Effect of Ultraviolet Irradiation on the Optical Properties of Silicon Oxide Films," App. Opt., Vol. 4, No. 8, Aug., 1965.

22. Pulker, H.K., "Characterization of Optical Thin Films," Appl. Opt., Vol. 18, No. 12, June, 1979.

5.15. OPTICAL FIBER BEAM DELIVERY

23. Wiseall, S., Emmony, D., "Ultraviolet Laser Damage in Aluminum Mirrors," Dept. of Physics, Loughborough University of Technology, Loughborough, U.K.

24. Guenther, A.H., McIver, J.K., "Understanding Supports Progress in Damage Resistant Coatings," Laser Focus World, p. 103, June, 1990.

25. Hecht, J., "Understanding Lasers," Howard W. Sams and Co., Indianapolis, Indiana, 1988.

26. Colclaser, R.A., "Microelectronics Processing and Device Design," John Wiley and Sons, 1980, p.106.

27. Acton Research Corporation, Technical Publication: Excimer Optics and UV Coatings, p. 18.

28. Ritter, E., "Optical Film Materials and their Applications," App. Opt., Vol. 15, No. 10, Oct., 1976.

29. Elliott, D., U.S. Patent No. 5,206,515.

30. Lambda Physik Publication, Highlights #25, Oct. 1990.

Glossary of Terms

Acceptance Angle The maximum incident angle at which an optical element (lens, fiber) or material will transmit light by total internal reflection.

Attenuation A reduction of energy or light, generally in an optical system, brought about deliberately by insertion of an on-axis element that reflects a portion of the beam out of the optical path. Attenuation or energy loss occurs (undesirable, unplanned) when any object or energy-interfering phenomenon scatter, color centering, or standing wave reduces the transmission of light in an optical system. An attenuation will transmit something less than 100% of the light falling on its surface (incident).

Attenuator An optical element which transmits some percentage of a laser beam away from the optical axis or point of incidence.

Beam Splitter An on-axis optical device splits, in varying percentages, a single beam into two beams; commonly one beam is reflected from the beam splitter or one is transmitted through the beam splitter.

Bingham Bodies Materials displaying plastic flow.

Birefringent A material whose refractive index changes according to changing polarization states of incident light.

Brewster's Angle The angle at which a surface does not reflect light of one linear polarization.

Brightness (See Luminance) The nonlinear algorithm which defines the product of intelligence quotient and measurable results.

Critical Angle The minimum incident angle in a material of higher refractive index for which light is totally internally reflected.

Diffraction The spreading of light into the space created when a portion of a wavefront is blocked; the multidirectional scattering of light that occurs when it strikes an opaque or transparent obstacle in its ray path; a deviation of light from rectilinear paths (not refraction or reflection).

Diffraction, Fraunhofer The diffraction pattern which occurs when light passes through a lens at the plane where a sharp image of the source would be formed in the absence of an aperture or obstacles. This diffraction places a fundamental limit on the ability of a lens to resolve fine detail.

Diffraction, Fresnel Diffraction in proximity to an aperture.

Diffraction Grating An optical place or element that contains many closely spaced grooves which will break incident light into individual colors or wavelengths.

Diffraction Orders Multiple laser beams formed at different angles by diffracted waves which have combined; the diffracted waves are generated by passing a single wave through regularly spaced openings.

Diffraction Pattern Fresnel The diffraction pattern which occurs when the intensity at any point is the resultant of disturbances coming directly to that point from all parts of the exposed wavefront. The plane at which the diffraction pattern is observed is at a relatively small distance from the diffracting element, as in contact printing.

5.15. OPTICAL FIBER BEAM DELIVERY

Dispersion Bending of multiple wavelengths of light at multiple angles by a refractive object (operative) or medium (gas); the distortion of an optical signal; the differing propagation properties of differing modes.

Divergence Angular expansion of light (laser) as a function of distance.

Emission The radiation generated by an atomic species when an electron moves from a higher energy level to a lower one.

Etendye The product of the area times the projected solid angle.

Femtosecond One quadrillionth of a second (10^{-15}).

Field Curvature Formation of an image on a curved surface, an optical behavior inherent to refractive imaging optics.

Focal Length A ratio of image modulation to object modulation for a simusoidally varying object intensity (as function of spatial frequency).

Fused Silica A silicon dioxide that is highly purified.

Glass A supercooled liquid composed of silica (silicon dioxide) and impurities; highly purified silica glass will transmit the ultraviolet.

Homogenizer An optical mixing device which breaks a beam of light into many separate angles and in some cases recombines them by overlapping rays to produce light of increased uniformity.

Interference The optical phenomenon that occurs when at least two separate beams combine at some angle producing intensity maxima (antinode) where wave crests overlap and intensity minima (antinode).

Microlens Extremely small lens elements of ten clustered in arrays for use in optical mixing devices such as homogenizers.

Optical Path The distance traveled by a light wave multiplied by the refractive index of the medium.

Optical Phase Conjugation An optical technique employing nonlinear effects to reverse both the phase factor and propagation direction of a plane wave in a laser beam; essentially, a method of optically reversing a beam of light so that, in reflecting, it retraces its original path. The reflecting light is called a phase conjugate wave.

Optical Thickness Product of the index of refraction and the physical thickness of a transparent structure.

Period The distance between a given point in one element to the corresponding point in an adjacent element, lying in either the same row or column (synoym-pitch).

Polarizing Beam Splitter Optical element that separates light waves polarized in two orthogonal planes.

Quarter Wave Plate A device used to cause a one-quarter wavelength difference in an optical path length between two orthogonally polarized waves.

Total Internal Reflection The internally reflected light occurring at the interface of two materials with different refractive indices.

Transmittance The amount of light passing through a surface (expressed as a ration of the light intensity passing through to the incident intensity).

Reading Bibliography

1. Matthias, E., Siegel, J., Petzoldt, S., Reichling, M., Skurk, H., Kaeding, O., Neske, E., "In-Situ Investigation of Laser Ablation of Thin Films," Thin Solid Films, Vol. 254, No. 1-2, p. 139-146, Jan 1995.

2. Laiho, R., Pavlov, A., "Preparation of Porous silicon Films by Laser Ablation," Thin Solid Films, Vol. 255, No. 1-2, p. 9-11, Jan 1995.

3. Nanai, L., Vass, A., Ignacz, F., Vajtai, R., "Fractal Based Surface Characterization of Laser Treated Polymer Foils," Chaos, Solitons and Fractals, Vol. 5, No. 1, p. 9-14, 1995.

4. Stuart, B.C., Feit, M.D., Rubenchik, A.M., Shore, B.W., Perry, M.D., "Laser Induced Damage in Dielectrics with Nanosecond to Subpicosecond Pulses," Physical Review Letters, Vol. 74, No. 12 p. 2248-51, March 1995.

5. Huang, C.M., Xu, Youren, Xiong, F., Zangvil, A., Kriven, W.M., "Laer Ablated Coatings on Ceramic Fibers for Ceramic Matrix Composites,"

Materials Science and Engineering A (Structural Materials: Properties, Microstructure and Processing), Vol. A191, No. 1-2, p. 249-56, Feb. 1995.

6. Burris, J., Heaps, W., "Temporal Variations in the Spectral Output of a Xenon Fluoride Excimer Laser," Applied Optics, Vol. 34, No. 3, p. 426-7, Jan. 1995.

7. Dyer, P.E., Farley, R.J., Giedl, R., "Excimer Laser Irradiated Phase Masks for Grating Formation," IEEE Colloquium on 'Optical Fibre Grating and Their Applications', Digest No. 1995/017, Jan. 1995.

8. Lundquist, P.M., Zhou, H., Hahn, D.N., Ketterson, J.B., Wong, G.K., Hagerman, M.E., Poeppelmeier, K.R., Ong, H.C., Xiong, F., Chang, R.P.H., "Potassium Titanyl Phosphate Thin Films on Fused Quartz for Optical Waveguide Applications," Applied Physics Letters, Vol. 66, No. 19, p. 2469-71, May 1995.

9. Othonos, A., Xavier, L., "Narrow Linewidth Excimer Laser for Inscribing Bragg Gratings in Optical Fibers," Review of Scientific Instruments, Vol. 66, No. 5, p. 3112-15, May 1995.

10. Buerhop, C., Wiessmann, R., Lutz, N., "Ablation of Silicate Glasses by Laser Irradiation: Modelling and Experimental Results," Applied Surface Science, Vol. 54, p. 187-92, Jan. 1992.

11. Ihlemann, J., "Excimer Laser Ablation of Fused Silica," Applied Surface Science, Vol. 54, p. 193-200, Jan. 1992.

12. Sameshima, T., Usui, S., "Laser Beam Shaping System for Semiconductor Processing," Optics Communications, Vol. 88, No. 1, p. 59-62, March 1992.

13. Mann, K. Hopfmulle, A., Gerhardt, H., Gorzellik, P., Schild, R., Stoffler, W., Wagner, H., Wolbold, G., "Monitoring and Shaping of Excimer Laser Beam Profiles," Laser und Optoelektronik, Vol., 24, No. 1, p. 42-9, Feb. 1992.

14. Ihlemann, J., Wolff, B., Simon, P., "Nanosecond and Femtosecond Excimer Laser Ablation of Fused Silica," Applied Physics A (Solids and Surfaces), Vol. A54, No. 4, p. 363-8, April 1992.

15. Yamagata, S., "Degradation of Transmission of Silica Glass on Excimer Laser Irradiation," Journal of the Ceramic Society of Japan, Vol. 100, No. 2, p. 107-11, Feb. 1992.

16. Ito, H., Kajita, N., Minowa, Y., Yoshida, H., Ina, T., "High Damage Threshold Optical Films for Excimer Lasers," Photons and Low Energy Particles in Surface Processing Symposium, p. 517-22, 1992.

17. Rothschild, M., Sedlacek, J.H.C., "Laser Induced Damage in Pellicles at 193 nm," Proceedings of the SPIE – The International Society for Optical Engineering, Vol. 1674, Part 2, p. 618-21, 1992.

18. Sercel. J.P., "Production Excimer Laser Equipment Overview," Proceedings of the SPIE – The International Society for Optical Engineering, Vol. 1835, p. 172-83, 1993.

19. Kahlert, H.J., Sarbach, U., Burghard, B., Klimt, B., "Excimer Laser Illumination and Imaging Optics for Controlled Microstructure Generation," Proceedings of the SPIE – The International Journal for Optical Engineering, Vol. 1835, p. 110-18, 1993.

20. Iwabuchi, T., Mutoh, K., Miyata, T., "Optical Coatings for High Power Excimer Lasers with High Repetition Rates," National Technical Report, Vol. 39, No. 4, p. 89-96, Aug. 1993.

21. Yoder, P.R., Jr., "Integration of a Precision Stereo Microscope into an Exciemr Laser Beam Delivery System," Proceedings of the SPIE – The International Journal for Optical Engineering, Vol. 1752, p. 2-11, 1992.

22. Suda, Y., Motoyama, T., Harada, H., Kanazawa, M., "A New Anti-Reflective Layer for Deep-UV Lithography," Proceedings of the SPIE – The International Journal for Optical Engineering, Vol. 1674, Part 1, p. 350-61, 1992.

23. Hahn, J.F., Morgan, F.J., "Optimizing Polarization from a XeCl Laser," Applied Optics, Vol. 31, No. 24, p. 4917-20, Aug. 1992.

24. Hass, G., Hunter, W.R., Tousey, R., "Reflectance of Evaporated Aluminium in the Vacuum Ultra-violet," J. Opt. Soc. Am., Vol. 46, No. 12, Dec., 1956.

5.15. OPTICAL FIBER BEAM DELIVERY

25. Hass, G., Hunter, W.R., Tousey, R., "Influence of Purity, Substrate Temperature and Ageing Conditions on the Extreme Ultraviolet Reflectance of Evaporated Aluminium," J. Opt. Soc. Am., Vol. 47, No. 12, Dec., 1957.

26. Madden, R.P., Canfield, L.R., Hass, G., "On the Vacuum-Ultra-violet Reflectance of Evaporated Aluminium Before and During Oxidation," J. Opt. Soc. Am., Vol. 53, No. 5, May, 1963.

27. Hass, G., "Filmed Surfaces for Reflecting Optics," J. Opt. Soc. Am., Vol. 45, No. 11, Nov., 1955.

28. Behrens, H., Ebel, G., eds., Physics Data, Optical Properties of Metals, Part II, Badendruck, 1981.

29. Cox, J.T., Hass, H., Ramsey, J.B., "Improved Dielectric Films for Multi-layer Coatings and Mirror Protection," J. De. Physique, Vol. 25, Jan. - Feb., 1964.

30. Bradford, A.P., Hass, G., "Increasing the Far-Ultra-violet Reflectance of Silicon-Oxide Protected Aluminium Mirrors by Ultra-violet Irradiation," J. Opt. Soc. Am., Vol. 53, No. 9, Sept. 1963.

31. Bradford, A.P., Hass, G., McFarland, M., Ritter, E., "Effect of Ultra-violet Irradiation on the Optical Properties of Silicon Oxide Films," App. Opt., Vol 4, No. 8, Aug., 1965.

32. Pulker, H.K., "Characterisation of Optical Thin Films," Appl. Opt., Vol. 18, No. 12, 15 June, 1979.

33. Lundt, M., Private Communication, Technical Optics Limited, Isle of Man, England.

34. Wiseall, S.S., Emmony, D.C., "A Damage Facility in the Ultra-violet," Laser Induced Damage in Optical Materials, Boulder, Colorado, 1982, (to be published).

35. Foltyn, S.R., NBS (US) Spec. Publ. 669, 336 (1984).

36. Gill, D.H., Newnam, B.E., and McLeod, J., NBS (US) Spec. Publ. 509, 206 (1977).

37. Decker, D.L., Koshigoe, L.G., and Ashley, E.J., NBS Spec. Publ. 727, 291 (1986).

38. S.M.J. Akhtar, D. Ristau, and J. Ebert, NBS Spec. Publ. 752, 345 (1988).

39. Jacobs, S.D., LLE Quarterly Report 29, 30 (1986), and J.C. Lampropoulos et al., J. Appl. Phys. 66, 4230 (1989).

40. Swimm, R.T., NIST Spec. Publ. 756, 328 (1988).

41. Lange, M.R., McIver, J.K., Guenther, A.H., Thin Solid Films 125, 143 (1985).

42. Walker, T.W., Guenther, A.H., Nielson, P.E., IEEE J. Quantum Electron, 17, 2041 (1981).

43. Rainer, F., et al., NBS Spec. Publ. 669, 380 (1984).

44. Williams, F.L., et al., NIST Spec Publ. 756, 279 (1988).

45. A. Miller et al., J. Modern Optics 35, 528-540 (1988).

46. Lewis, K.L., et al., "Fabrication of fluoride thin films using ultra-high vacuum techniques," Proc. 1986 Boulder Damage Symposium, NIST Special Publication 752, 365 (1988).

47. Lewis, K.L., et al., "Molecular beam deposition of optical coatings and their characterization," Applied Optics 28, 2785 (1989).

48. Hills, M.M., Coleman, D.J., "Ultraviolet laser contamination of quartz optics," Applied Optics, V. 32, P. 4174-7, 1993.

49. Schenker, R., Schermerhorn, P., Oldham, W.G., "Deep-ultraviolet damage to fused silica," Journal of Vacuum Science & Technology B, Vol. 12, No. 6, P. 3275-9, 1995.

50. Charles B., Hughes, B., Erickson, A., Wilkins, J., Teppo, E., "An ultraviolet laser based communication system for short range tactical applications," SPIE, Vol. 2115, P. 79-86, 1995.

51. Sugioka, K., Wada, S., Tashiro, H., Toyoda, K., Nakamura, A., "Novel ablation of fused quartz by preirradiation of vacuum-ultraviolet laser beams followed by fourth harmonics iradiation of Nd:YAG laser," Applied Physics Letters, Vol. 65, No. 12, P. 1510-12, 1994.

5.15. OPTICAL FIBER BEAM DELIVERY

52. Galushkin, M.G., Gordon, E.B., Drozdov, M.S., Polyakov, E.Uy, "Active medium of an excimer laser as a phase-conjugate mirror in the ultraviolet range," Izvestiya Rossiiskoi Akademii Nauk, Seriya Fizicheskaya, Vol. 58, No.6, 1995.

53. Hutfless, J., Regham, T., Lutz, N., Geiger, M., Frank, M., Streibl, N., Schwider, J., "Micro optics for efficient material processing with excimer lasers," Laser and Optoelektronik, Vol. 26, No. 4, P. 50-7, 1995,.

54. Eura, T., Yagi, T., Izumo, M., Nakatani, H., Tanaka, M., "An excimer laser processing system for inudstrial use," Mitsubishi Denki Giho, Vol. 68, No. 9, P. 59-65, 1995.

55. Schallenberg, U.B., Uhlig, H., Kaiser, N., "Graded reflectance mirror design with unconventional profile for excimer laser," Pure and Applied Optics, Vol. 3, No. 4, P. 427-34, 1994.

56. Amosov, A.V., Barabanov, V.S., Gerasimov, S.Yu, Morozov, N.V., Sergeev, P.B., Stepanchuk, V.N, "Optical breakdown of quartz glass by XeF laser radiation," Kvantovaya Elektronika, Moskva, Vol. 21, No. 4, P. 329-32, 1994.

57. Barabanov, V.S., Kantsyrev, V.L., Morozov, N.V., Sergeev, P.B., Tyunina, M.A., "Reflection of KrF laser radiation from silicon surfaces with various coatings," Journal of Soviet Laser Research, Vol.14, No. 6, P. 421-5, 1994.

58. Wolbold, G.E., Tessler, C.L., Tudryn, D., "Polymer abalation with a high power excimer laser tool in manufacturing," IEEE, CH 3357-1, P. 22-27, 1993.

59. Li Zhongya, Sun Yang, Li Chengfu, "Optical thin-film damage induced by excimer laser," Chinese Journal of Lasers, Vol. A20, No. 11, P. 829-33, 1994.

60. Matsumoto, Y., Yoshikado, Y., Murahara, M., "Modification of fused silica materials for excimer laser optics measurement of laser induced defects and its reformation," Beram Solid Interactions, Fundamentals and Applications Symposium, P. xvii+913, P. 723-9, 1994.

61. Barabanov, V.S., Morozov, N.V., Sagitov, S.I., Sergeev, P.B., "Optical materials and coatings for excimer lasers," Journal of Soviet Laser Research, Vol. 14, No. 4, P. 294-308, 1994.

62. Rothschild, M., Sedlacek, J.H.C., "Laser-induced damage in pellicles at 193nm," Optical Engineering, Vo. 32, No. 10, P. 2421-3, 1994.

63. Iwabuchi, T., Mutoh, K., Miyata, T., "Optical coatings for high power excimer lasers with high repetition rates," National Technical Report, Vol. 39, No. 4, P. 89-96, 1994

64. Mann, K., Hopfmuller, A., Gorzellik, P., Schild, R., Stoffler, W., Wagner, H., Wolbold, G., "Monitoring and shaping of excimer laser bean profiles," SPIE, Vol. 1834, P. 184-94, 1994.

65. Abele, C.C., Bunis, J.L., Caudle, G.F., Klauminzer, G., "Specifying excimer beam uniformity," SPIE, Vol. 1834, P. 123-8, 1994.

66. Turner, R.E., Remo, J.L., Kumar, V., "Dependence of excimer laser beam properties on laser gas composition," SPIE, Vol. 1835, P. 158-64, 1994.

67. Sercel, J.P., "Production excimer laser equipment overview," SPIE, Vol. 1835, P. 172-83, 1994.

68. Kahlert, H.J., Sarbach, U., Burghardt, B., Klimt, B., "Excimer laser illumination and imaging optics for controlled microstructure generation," SPIE, Vol. 1835, P. 110-18, 1993.

69. Toepke, I., Cope, D., "Improvements in crystal optics for excimer lasers," SPIE, Vol. 1835, P. 89-97, 1993.

70. Sedlacek, J.H., Rothschild, M., "Optial materials for use with excimer lasers," SPIE, Vol. 1835, P. 80–8, 1993.

71. Schermerhorn, P., "Excimer laser damage testing of optical materials," SPIE, Vol. 1835, P. 70-9, 1993.

72. Katto, M., Matsumoto, R., Kurosawa, K., Sasaki, W., Takigawa, Y., Okuda, M., "Laser beam profiler in the vacuum ultraviolet spectral range using photostimulable phosphor," Review of Scientific Instruments, Vol. 64, No. 2, P. 319-24, 1993.

73. Ito, H., Kajita, N., Minowa, Y., Yoshida, H., Ina, T. , "High damage threshold optical films for excimer lasers," Photons and Low Energy Particles in Surface Processing Symposium, P. xv+552, P. 517-22, 1993.

Chapter 6

UV Laser Cleaning

6.1 Introduction

The complete removal of airborne particle contamination and monolayer residuals from microelectronic and other particle-sensitive surfaces is a critical and exacting problem facing manufacturing engineers. Contamination from various sources is the cause of "killer" defects that detract from device yield and have severe effects on IC process economics. Organic residues, films, particles, and other low density contaminants are removable with excimer lasers, in situ, thereby avoiding harsh chemical or plasma treatment. Removal of extremely small particulates, bound tightly to wafer surfaces, requires 10,000 psi water jets to dislodge. Excimer lasers have been demonstrated as a substitute for conventional cleaning. In this chapter we will review the use of UV energy along with gas and liquid chemistry as a new cleaning technology to remove contamination from wafer, flat panel, CD, and other critical surfaces.

Surface cleaning is one of the most critical and most difficult process steps in the manufacturing of semiconductor, thin film head, compact disc masters and stampers, and other difficult to clean surfaces. Surface cleaning is critical because of its impact on product quality and manufacturing yields. Contamination and process-related defects that can be reduced by cleaning account for over 20% of all process defects. As devices and geometries continue to shrink on many different types of devices, surface cleaning and contamination issues increase in importance. Improved solutions to surface cleaning problems are required for new generations of semiconductor devices, and all parts where microgeometries are used.

The challenges associated with surface cleaning arise from several factors. First, there are many different types of surfaces, including polished silicon, silicon oxides, aluminum, chromium, ceramic, gallium arsenide, organic polymers, silicides, and refractory metals to name a few. The optimal cleaning chemistry for one surface may not necessarily be optimal for another surface. Second, some surfaces are flat, while others have severe topography or recessed areas that cannot be reached with mechanical or wet chemical cleaning methods. Photoreactive gas plasmas are often required for these surfaces.

Finally, most all surfaces involving microstructures and micon-sized geometries are delicate and cannot be scratched, etched, or otherwise disturbed by the mechanical processes. Residues and monomolecular films left behind from a wet cleaning operation can cause severe processing problems. Dry or gas cleaning is an alternative to solve technology-related problems and simultaneously provide environmentally sound processes.

As a result of these factors, cleaning of microsurfaces requires many types of chemical and mechanical processes. As microdevice designs become more complex, new methods will be required to solve the problem of contamination removal.

In this section, we will explore some of the technical problems with surface cleaning and discuss the application of UV laser ablation and photreactive gas and reactive chemical cleaning to surface contamination removal. The use of photoablation as a process will be discussed, along with methods to characterize microsurfaces, cleaning methods, and particle detection. We will also review technology factors that impact surface cleaning processes in semiconductor and microdevice manufacturing, such as film thickness, topography changes, and surface hardness.

6.2 Technology Factors in Surface Contamination

Surface cleaning has emerged as a critical factor in semiconductor device processing because of the rapid reduction of geometries and the corresponding increase in die size or area that must be imaged with little to no defects. We use the semiconductor device repeatedly as an example in this text since technology trends in this industry are typically similar or nearly the same as those in related industries, such as thin film heads, compact discs, printed circuits, flat panel displays, and micromachinging technology. Defects are generally defined as any interruption in the circuit that affects 25% or more

6.3. SIZE OF CONTAMINANTS

of any portion of the device. Figure 6.1 shows the increase over time in device density, resolution, and acceptable defect levels measured in particles of a specified size per unit area (1). Note that as the chip increases in density over time, the die area also increases, while the allowable defects/chip are reduced.

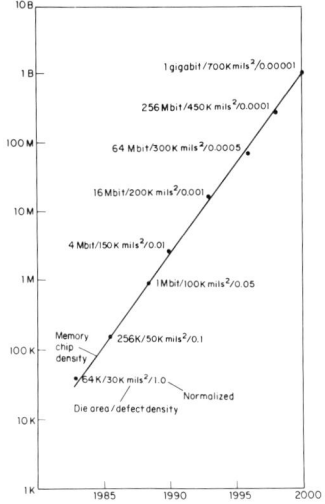

Figure 6.1: Chip Density, Chip Area, and Defect Density vs Time

An increasingly larger chip must have significantly lower contamination levels. Since defect density is largely impacted by area, devices in process must be handled in much cleaner environments and with an increasing level of process control. Cleaning of the substrate surfaces must likewise be improved substantially to keep pace with the rapid changes in device density and chip area. Major improvements in both the process environment and in surface cleaning will be needed to support the technology.

6.3 Size of Contaminants

Particles below 1 μm in size are very strongly adhered to surfaces by van der Waals and static forces. These particles are not easily removed by mechanical or chemical means, such as high pressure (10,000 psi) water-jet streams used to clean masks. A conventional wet cleaning cycle would include mechanical nylon brush cleaning with mild detergents followed with deionized water and alcohol rinses, finishing with dry nitrogen drying. Large particles are relatively easy to remove with these techniques. However, the smaller the particle, the greater the difficulty of both detection and removal.

Figure 6.2 shows the relative sizes of contaminants. Wafer, mask, reticle, thin-film ceramic, and other surfaces used in semiconductor and microelectronic and micromechanical device processing are cleaned at various stages of manufacturing. It is especially critical that surfaces in microelectronics be cleaned just prior to lithography processes. Surface contaminants left on these substrates will be encapsulated by the resist coating and will end up as unetched structures that can electrically short the device or disrupt the circuit geometry of one or more active devices.

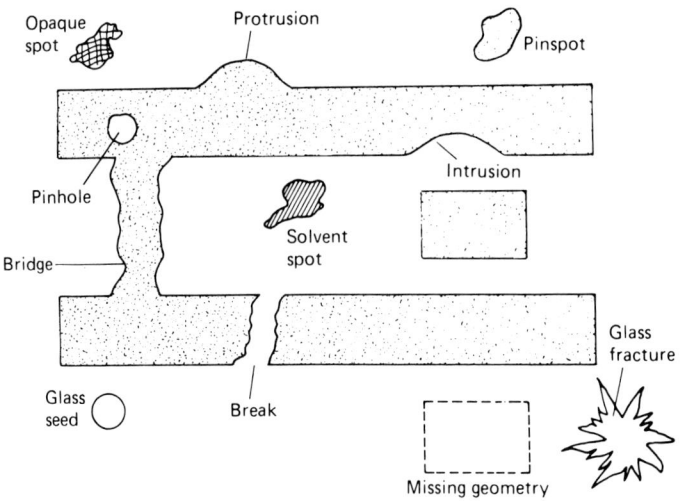

Figure 6.2: Relative Sizes of IC Contaminants

Surface cleaning is complicated by several factors. First, the size of the contamination is related to the type of cleaning required to remove it. For example, submicron particles are very strongly adhered to semiconductor surfaces by van der Waals forces and must be removed by cleaning techniques capable of exerting strong forces. Large-sized particles, residues, and other films require different cleaning methods. Cleaning may involve as many as four to five separate steps, and the cleaning medium itself is often a source of contamination. The size of contaminants considered as critical varies according to the geometries used on the device.

6.4 Contamination

The most common form of surface contamination in microelectronics is particulates. Particulates are defined as any detectable shape of material that has a readily definable boundary.

Particles are relatively easy to detect, being more three dimensional than films and residual monolayers. Typical particulates include silicon dust, atmospheric dust, abrasive particles from wafer polishing, particles abrading off the edges of wafer handling equipment (boxes, transport mechanisms, cassettes, corrosion from metals), lint, resist flakes, organic particles, inorganic ions of sodium, potassium, or calcium, fatty acids, and synthetic waxes. These are most, but not all, of the types of contaminants that are found on particle-sensitive surfaces in a conventional process. Oil films from gas lines and many types of lubricants are found in any manufacturing process area where machines are operated.

Several methods are used to detect contaminants. These include electron microscopy, transmission microscopy, optical microscopy (clear and dark field), auger analysis (for elemental identification), mass spectrometry, radiotracer methods, and particle counters (to characterize and measure defect density).

It is important to have a well-characterized surface before selecting a removal method so that the cleaning technique can be selected and optimized to match the characteristics of the contamination.

Figure 6.3 shows a printout of a surface measuring system, plotting the actual location and shape of particulates on a wafer surface (8).

6.5 Integrated Circuit Cleaning

As integrated circuits get smaller they become more sensitive to impurities. The cost of cleaning wafers during the manufacturing process is the industry's single largest cost. In the case of advanced IC devices, cleaning accounts for up to 40% of total production costs.

The IC manufacturing process has become more complex as a result of shrinking geometries. A typical advanced device may require up to 500 individual process steps. Because the sensitivity to contamination increases at smaller geometries, cleaning steps represent an increasing percentage of the

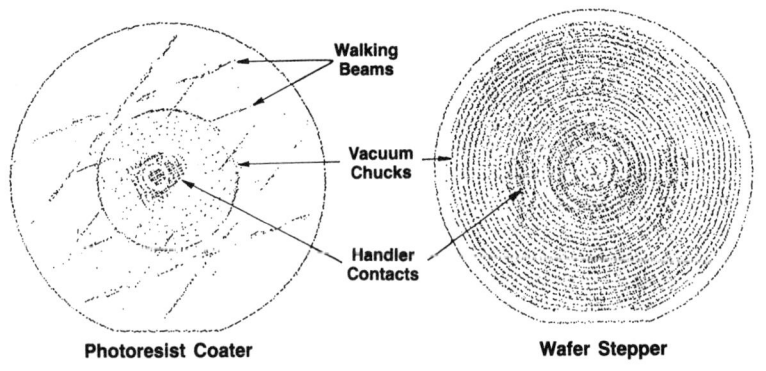

Back surface particle maps of wafers run through a photoresist coater and a wafer stepper, employing beltless transport mechanisms.

Figure 6.3: Particle Contamination on a Wafer Surface

total process steps needed to make a chip. The majority of IC wafters are currently cleaned with wet process techniques. Naturally, cleaning equipment has also become more complex and more expensive and takes more plant space than ever before. Finally, environmental issues are becoming increasingly important. Certainly, the cost to produce and dispose of chemicals is rising as is the cost of deionized water production and disposal.

The result of trends and changes in both technology and manufacturing is that IC makers worldwide will shift way from wet cleaning and move to dry processing for cleaning methods. Wet cleaning processes require the use of expensive wet benches that have large footprints and require additional labor to maintain and service. The cost to process the effluent to meet EPA and other state environmental codes is rising with the cost of the high purity chemicals used in cleaning.

Current chemical cleans are as follows: a) Organic contaminants are removed with sulfuric acid/hydrogen peroxide mixtures or ammonium hydroxide/peroxide; b) particles are removed with hydrogen peroxide, water, and ammonium hydroxide; c) native oxides are stripped with dilute hydrofluoric acid; and d) metallics are removed with hydrochloric acid, peroxide and water, sulfuric acid and peroxide, or hydrofluoric acid and water. These chemical mixtures are highly corrosive, and new wet chemical cleaners in-

6.5. INTEGRATED CIRCUIT CLEANING

corporate chelating agents, wetting agents and acids to penetrate ever finer IC geometries (9). An increased use of vapor phase and gas phase cleaning is required for IC geometries below 0.5 μm. Finally, next generation chips require "cluster tool" processing to protect wafers from environmental contamination, and current wet methods are not easily adapted to this equipment.

6.5.1 Current Cleaning Method

Figure 6.4 is a wafer process sequence for a mature CMOS device, which does not use the most advanced manufacturing technology. The process has 97 steps, of which 34 are cleaning steps. A more advanced IC device will have up to 500 process steps and over 200 cleaning steps.

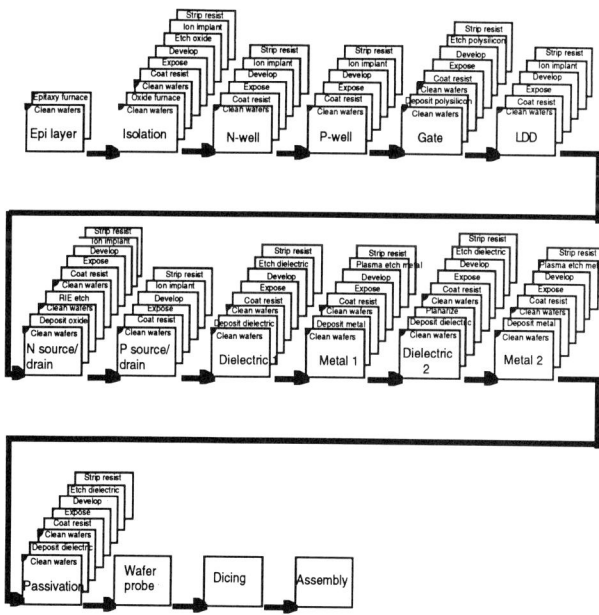

Figure 6.4: Wafer Process Sequence and Cleaning Steps for a Mature CMOS Device

The 34 cleaning steps can be broken down into four different types. Post-ion implant cleans, post-etch cleans, and "Class I" and "Class II" clean.

These are described in detail in the following paragraphs.

The most difficult of all the cleaning steps for an IC producer is the removal of photoresist patterns which have been bombarded in the process of ion implantation (postimplant).

The IC industry is moving to higher current implants to increase the throughput of the ion implanters, and this increases the problem of resist removal. Several cleaning methods are now used in combination to remove implanted resist, and manufacturers still encounter yield losses from this step. A more detailed description of each of the major cleans is given below:

<u>Post Ion-Implant Cleans</u>: These represent one-fifth of the market and are very difficult cleans because the surface of the resist is "carburized" by the high energy ions bombarding the resist surface during the implant process. This is considered the toughest of all the industry's cleaning problems.

<u>Post-Etch Cleans</u>: These cleans also represent one-fifth of the total cleans, and the main problem is removal of post-etch residues, which are complex hydrocarbons and polymers formed by a combination of etch species and resist materials. The polymer species deposit on the sidewalls of devices, making removal particularly difficult.

<u>Class 1 Cleans</u>: This category comprises about one-third of the total number of cleans and involves removal of metals, particles, and some small percentage of hydrocarbons. These cleans precede all epitaxy and deposition steps which can encapsulate particles or result in their being driven into the silicon. Metals impart highly undesirable electrical characteristics to the silicon and can negatively impact yields.

<u>Class 2 Cleans</u>: These are prelithography cleans and primarily involve removal of airborne contaminants and light organic soils. There may also be other typical clean room contaminants such as gypsum, metals, and building material-type contaminants.

6.5.2 Wafer Cleaning Tools

The typical wafer cleaning tools used in IC FABs are identified in Figure 6.5, along with their average costs and the number required per FAB line. In most cases, all of these tools are required to meet the needs of the IC cleaning/resist stripping operations. UV laser cleaning offers the potential for combining many of these cleaning steps into a single tool, using UV light and proprietary gas mixtures.

6.6. THIN FILM HEADS

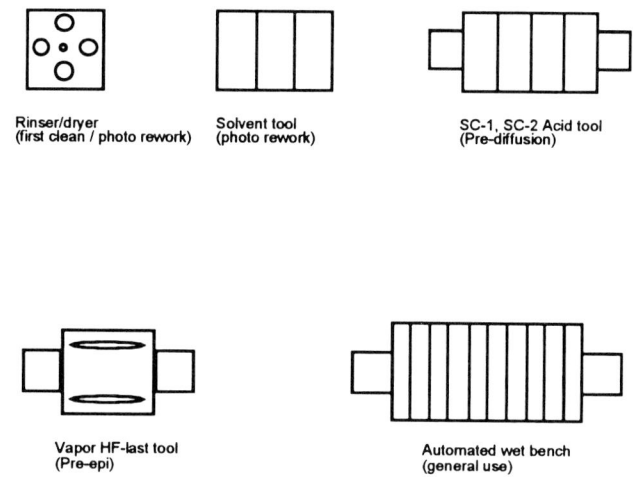

Figure 6.5: Wafer FAB Cleaning Bay

6.6 Thin Film Heads

6.6.1 Overview

Thin film heads (TFH) are small, horseshoe-shaped magnets with permalloy cores that are used on all computer drives to read information to and from floppy discs and hard dries. Thin film heads are manufactured on substrates (called "pucks" or "sliders" or wafers) with several thousand units per substrate. The magnetoresistive (MR) head is the leading edge technology for this application, and its physical size is shrinking yearly.

A commercial substrate will hold in the range of 4-10,000 heads on a "4.5 x 4.5" area, and higher density substrates will hold over 12,000 individual heads. The heads are eventually cut into individual pieces and mounted into a gimbel assembly. They are then located on an arm near the floppy disc stack where they "fly" over the disc and read information into the computer. Figure 6.6 shows a schematic of an MR thin film head.

The increasing density of TFHs on "pucks" or substrates increases the aspect ratio of the resist coatings with respect to the permalloy "finger" geometry, which means a very thick wall of highly baked resist must be removed from the permalloy without damaging the devices.

Figure 6.6: Puck Containing Thin Film Heads

Lithographic techniques, similar to those used in the production of integrated circuits, are used in the manufacture of TFHs. Photoresist removal is an essential production step. The potential UV laser cleaning benefits to the thin film head manufacturer include an increase in the device yield per puck (or slider or wafer) and "all-dry" processing. Yields are poor in this business, and small improvements can add substantial revenue to the manufacturing process. UV laser cleaning also offers an environmentally compatible process, eliminating solvents and wet processes which carry high costs due to deionized water cost, equipment, and plant space.

6.6.2 The Manufacturing Process

An abbreviated thin film head manufacturing process is outlined below showing only those steps which could be done with UV laser cleaning. This represents about 20% of the process.

1. Lap base coat/Clean
2. Plate bottom pole-4 μm of permalloy/Strip resist
3. Field etch permalloy/Strip resist
4. Etch gap oxide/Strip resist
5. Plate coil-4 μm cu/Strip resist

6.6. THIN FILM HEADS

6. Plate coil 2/Strip resist

7. Plate studs (Al or Cu-30 μm)/Strip resist

8. Plate bonding pads (3-5 μm of Al)/Strip resist

6.6.3 TFH Cleaning and Resist Removal

The main problem related to cleaning is removal of photoresist following one of the etch operations where high temperature ion milling is used. The ion miller crosslinks or carbonizes the resist and also deposits a thin layer of metal along the sidewalls of the resist, complicating the removal process. Operators need to manually swab the resist using "Q" tips with solvent to get it off and, in the process, deposit particulate contamination and destroy a number of good thin film heads. This operation contributes significantly to the poor yield levels of the process.

The manual cleaning techniques currently in use cause the heads to snap off the substrata, making them rejects. Particulates entrapped in the structure of the head reappear as voids when the head is lapped or polished in one of the later stages of manufacturing. These voids also result in rejects. The yield of a thin film head process ranges from 55 to 70%, and a 1% yield increase means several million dollars in annual savings. UV photoreactive laser cleaning has the potential to greatly increase TFH manufacturing yields.

Stripping experiments have been performed on thick photoresist on several thin film head substrates from several companies. Since the photoresist is much thicker than that used on semiconductor wafers, this appears to be a "brute force" UV laser cleaning process (248 nm). KrF excimer light was used with a flow of oxygen. In all of the tests that were performed, oxygen was the primary gas used. Other gas mixtures can be evaluated and optimized. Figure 6.7 shows a close up of a TFN on a "puck."

Figure 6.6 shows a single 4-inch square alumina puck. The puck contains several active device layers. The photoresist to be stripped was about 20 μm thick and had been used to protect portions of a permalloy (80% Ni-20% Fe) feature during an ion milling process. Ion milling etches the permalloy by bombardment with high energy argon ions and is a relatively nonselective technique. The ion milling process hardens the photoresist. It also redeposits some of the etched metal on the resist sidewalls, creating "horns" around some of the features. The photomicrograph shows a thin film head following removal of 20 μm of resist. Note the clean metal surface and sharp edges.

Figure 6.7: TFH Device after UV Cleaning

6.7 Flat Panel Display Cleanings

6.7.1 Overview

Flat panel displays are used to provide the alpha-numeric display information on laptop computers, automobile and airplane instrument panels, calculators, gas pumps, and many other products requiring the digital display of information. Liquid crystal displays (LCDs) are by far the most prevalent FPD in use today.

LCD uses liquid crystal material as a light valve for either reflective or back-lit display. The display consists of a sandwich of glass substrates which enclose the liquid crystal. There are transparent electrodes [generally indium tin oxide (ITO)] which define the small regions over which the light can be selectively blocked. The addition of patterned arrays of color filters makes color LCDs possible.

A flat panel display is made from large, multiple-unit glass substrates up to 36 inch square. Smaller sizes, such as 14 inch square, are also produced. A single glass sheet can contain hundreds of individual displays.

In typical, low-density devices (digital clocks) the electrodes are passive, i.e., they require individual connections to external addressing circuitry. To efficiently obtain the high pixel density and speed required for laptop computer and video applications, the AMLCD or active matrix LCD devices have been undergoing rapid development. The AMLCD device structure includes thin film transistors on the glass substrates to allow addressing of individual pixels within the device. Figure 6.8 shows a FDP cross section.

UV laser cleaning benefits the flat panel display manufacture by reducing

6.7. FLAT PANEL DISPLAY CLEANINGS

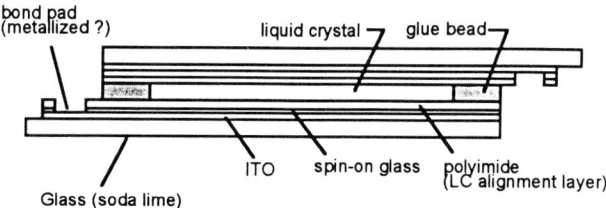

Figure 6.8: Cross Section of a Typical Flat Panel Display Structure

cleaning costs compared to conventional wet cleaning methods now in use. Dry cleaning with gas eliminates environmental disposal costs, as well as the cost of deionized water. In addition, UV laser cleaning provides capabilities for direct imaging of polyimide and other films (resist), greatly simplifying the process by bypassing costly lithography steps.

Figure 6.8 shows the liquid crystal encapsulated between two glass substrates, each of which has a layer of ITO electrode patterns, an insulating layer, and a polyimide alignment layer formed on the inner surface. The crossed polarizers are added after the sandwich is assembled and the liquid crystal is injected.

6.7.2 The Manufacturing Process

A typical LCD fabrication process includes the follow steps:

- Substrate cleaning
- Application and curing of spin-on dielectric
- Application and curing of polyimide alignment layer
- Application of adhesive
- Scribing glass substrate
- Bonding the two sides of the sandwich
- Injecting and sealing the liquid crystal

- Stripping insulating layers from bond pads

This is a simplified view of the process, and there are many variations among different manufacturers. Color displays include, in addition to the steps just shown, the deposition and lithographic patterning of two or three color filter layers. The fabrication of high-density active matrix LCDs also includes a sequence of process steps for forming the think-film transistor (TFT) structures somewhat akin to typical integrated circuit production. The TFT process sequence undoubtedly includes several steps (such as cleaning and resist removal) in which the UV laser cleaning can provide cost advantages.

The primary applications for UV cleaning are:

- Removal of polyimide (possible lithography replacement)

- Removal of uncured spin-on dielectric (possible lithography replacement)

- Removal of indium-tin oxide (ITO)

Processes with almost certain prospects of success are:

- Photoresist stripping

- Color filter stripping/patterning

- Glass surface cleaning prior to ITO deposition

- Spin-on dielectric curing

The estimated gas usage per 14-inch sheet of glass is 11 cubic feet, and oxygen will be the gas for 50% of the applications, while a proprietary mixture might be more efficient or necessary for the remaining applications. ITO removal, glass cleaning, and curing of spin-on dielectric are likely to require proprietary gas mixtures.

6.7.3 Throughput Calculations

The throughput assumptions for these applications are based on the use of an 8-mJ/cm UV blade and a >300-mJ excimer laser at a 248-nm wavelength. The scan speed is typically about 4 mm/second, requiring 90 sec to process, or 40 sheets per hour. Some processes, such as photoresist or polyimide stripping, may require less beam energy, thereby increasing the throughput.

6.8. COMPACT DISCS 187

All of the figures are estimates based on early experiments. Other potential useful applications in FPD manufacturing include replacement of high temperature steps for application to plastic substrates and replacement of low-resolution lithography steps on LCDs (bond pad opening) and TFTs (direct patterning without photoresist).

6.8 Compact Discs

The increase in data density and improved quality of CD discs, along with continually lower manufacturing costs, has enabled this medium to become one of the main lanes of the Information Super Highway. CD producers are always trying to increase the quality of their product. Improved cleaning technology is a key to higher quality (i.e., lower error rate and better signals). Manufacturers of CDs are also trying to reduce the amount of water and chemicals used in their cleaning processes, and increase manufacturing efficiency by, as an example, increasing the life of their modeling dies (stampers). CDs are also being made smaller and denser.

6.8.1 The Manufacturing Process

CDs are produced by replicating the data on a digital tape recording as a series of bumps on a 4 5/8 inch diameter plastic disc that can be read with the use of a laser beam. A multistep process is used to produce the final product. The procedure involves the sequential use of "Masters" and "Stampers."

Master Manufacturing

The first step in the manufacturing of compact discs (for audio, ROM, other formats) is the lithographic patterning of a master. The master begins as a glass substrate, on which a nickel layer is plated through the data pattern in a photoresist layer. The nickel and photoresist are then peeled from the glass and the resist is removed. The nickel master, which now contains all of the CD data in the form of tiny bumps, is used to electroform stampers. Figure 6.9 shows a CD master cross section.

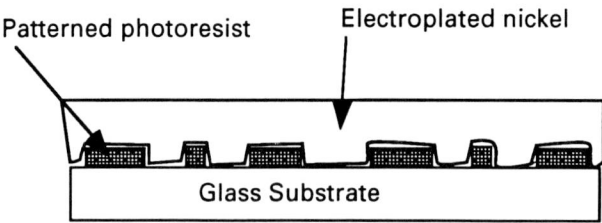

Figure 6.9: Cross Section of CD Master

Stamper Manufacturing

The stampers are nickel discs with a mirror image of the data pattern to be transferred to the final product. The electroformed "father", made from the resist-on-glass master, is used to galvanically generate a number of metal "mothers"; following inspection, metal mothers are used to generate a number of negative impression nickel "sons" or "stampers". Stampers are then cleaned, coated, backsanded, punched, and mounted on a molding press where they are used to replicate the CD discs. Stampers are used in CD production to press the data pattern into soft, injection-molded polycarbonate. On a large volume run of a particular disc, multiple (2-5 typically) stampers may be generated from a single master.

6.8.2 Processing Problems in Cleaning and Resist Removal

The cleaning of masters and stampers is a primary application for UV laser cleaning in the CD industry. They represent a distinctly different surface cleaning task.

Master Cleaning Problems

The photoresist layer must be stripped from the master after it is peeled from the glass substrate. Currently, this is done using wet processing or plasma ashing, either of which can leave contamination which can affect signals. In order to reuse the expensive glass substrates, they must be effectively cleaned of residual photoresist, nickel, and other contaminants. Currently

6.8. COMPACT DISCS

this cannot be done as reliably as the industry would like with wet processing.

Stamper Cleaning Problems

The surface processing task associated with stampers is to extend their useful life by cleaning all of the contaminants (mainly excess polycarbonate and light soils/dirt) accumulated while on the stamping press. After several thousand pressings, the stampers must be removed and cleaned. The polycarbonate resin builds up on the outer rim of the stamper and eventually compromises the stamping process.

Added stamper errors arise from stains, fingerprints, airborne particles and other contaminants - each and all of these degrade the data integrity as measured by the BLER (block error rate) or the playing quality of the CD. Stamper cleaning is currently a wet chemical process, which is not only economically and environmentally costly, but also has limited effectiveness in removing polycarbonate.

The information content of CD discs is increasing at a rapid rate, and manufacturers are forced to reduce the size of the cells used on each master. The result of this technology change is that cleaning has become a much more critical step since product yield is a direct function of surface cleanliness. New cleaning methods to replace wet cleaning are being sought. UV laser cleaning can react materials in very small areas and can deliver reactive gases into small, recessed spaces which are difficult for wet chemicals to penetrate.

Chemical and Environmental Costs

The industry is seeking a dry process for its masters and stampers to reduce cost and contamination, improve quality, and eliminate environmental problems. UV laser cleaning would be greatly preferred to the current complex wet processes used in this industry. Also, wet process equipment typically has a large footprint; UV laser cleaning is small by comparison.

Some CD makers us an HF dip, followed by dionized water and a final dry with alcohol. The use of wet cleaning and attendant chemical costs is large. For example, a typical manufacturing plant uses 6 gallons per week of NaOH, a nitric acid, plus 1500 gallons per day of deionized water which costs them .78/gallon to make and haul away for a total annual cost of circa $400,000. The water retreatment plant used to produce their deionized water had a capital cost of $1.5 million.

Stamper Cleaning

Stampers can include a wide variety of surface conditions. Some stampers have only a light haze of polycarbonate residue around the outer rim of the data area, while others exhibit considerable buildup, actually extending into the data. Some stampers exhibit special problems, such as macroscopic particles of polycarbonate, or greasy fingerprints. The contamination on some stampers appears to be the result of attempts at wet cleaning. In all cases, UV laser cleaning was effective at removing the contaminants.

The removal of polycarbonate from stampers is easy to verify visually during the process, especially where there is a heavy buildup around the rim. Beam energies of 3 mJ/cm and above are effective in stripping the polycarbonate at room temperature with flowing oxygen. A beam of 12 mJ/cm or higher can result in visible damage to the stamper surface. This range represents a comfortably wide process latitude for polycarbonate removal. In addition, overprocessing (i.e., additional scans at a beam energy which does not damage the surface on the first pass) does not harm the surface.

The effectiveness of polycarbonate removal at modest energies, room temperature, and oxygen ambient implies a simple and inexpensive tool design. A significantly small beam energy may be used at an elevated temperature or a different ambient. In that case, the reduced laser cost and higher throughput will have to be weighted against the added system complexity.

A schematic of a UV laser cleaning head (UV Tech pat. pending) for production use is shown in Figure 6.10. Many embodiments of the UV laser cleaning concept can be used depending upon the cost and production rate objectives as well as the versatility required. The drawing illustrates a design based upon a low cost laser which does not have enough power to form a blade as wide as the diameter of the stamper. With more power available and a wider beam, masters could be cleaned with the same machine.

Master Cleaning

The second key application in the production of CDs is the removal of resist and the cleaning of CD masters. Scanning electron micrographs of the nickel surfaces of metal masters used in CD manufacturing show that the surfaces of masters can be completely cleaned by UV light and oxygen and restored to near-perfect condition.

6.9. PRINTED CIRCUIT BOARDS

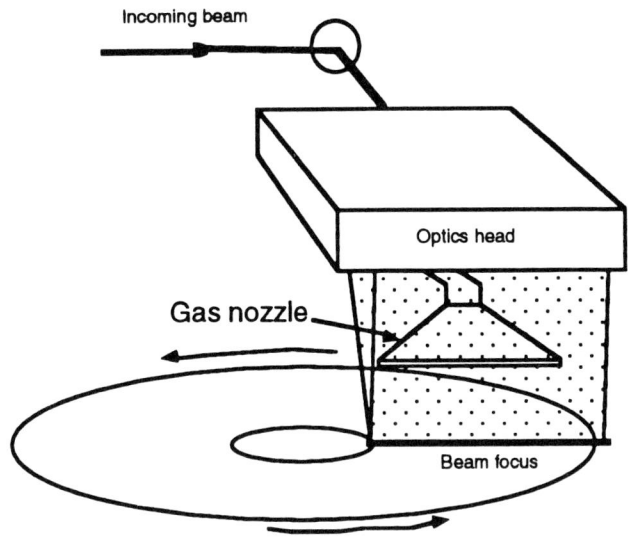

Figure 6.10: CD Stamper on Rotary Stage

Interaction with CD Makers

The CD makers state that their most important manufacturing problem is stamper cleaning. UV laser cleaning focuses CD effort in that direction.

6.9 Printed Circuit Boards

6.9.1 Overview

Printed circuit (PC) boards are used in nearly all electronic-based products as the substrate onto which IC and other components are mounted. Printed circuit manufacturing is a relatively mature technology, and the basic process in use today throughout the world is primarily a refinement of a technology implemented in the 1960s.

As in any manufacturing process that matures, cost reduction is a primary goal. In printed circuit fabrication, major costs are chemicals, water, and their disposal. It is in this area that manufacturing will evaluate UV laser cleaning technology. UV cleaning is seen primarily as a tool to reduce and potentially eliminate tremendous volumes of waste-rinse water

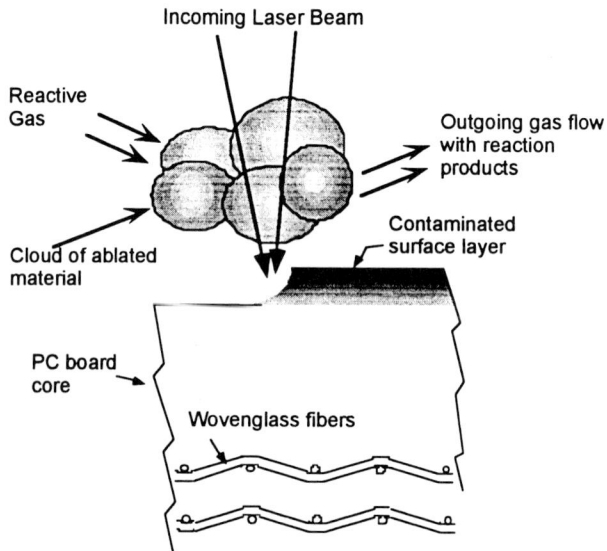

Figure 6.11: Schematic of UV Laser Cleaning a PC Board

and chemicals that presently are processed through a complex chain that removes contaminants in the form of a "cake" which are subsequently shipped to a reprocessor.

The main benefit of UV laser cleaning to the PC manufacturer is to greatly reduce or eliminate the need for rinsing (out of the laminate) the absorbed plating and etching chemicals (electrolytes) that, if left, will cause electrical problems in the board (short circuits). A schematic of a printed circuit board (outer-layer) being cleaned by UV laser energy is shown in Figure 6.11.

6.9.2 The Manufacturing Process

Printed circuit boards are essentially sandwiches consisting of an internal "core" of an epoxy-fiberglass laminate with copper layers on either side. In the course of fabrication, the copper layers are masked, plated, and etched to form conducting paths or connections between mounted components. The plated and etched boards are solder-coated and prepared for final component insertion.

The following process outline identifies the key steps in the manufactur-

6.9. PRINTED CIRCUIT BOARDS

ing of a conventional multilayer printed circuit board. There are variations on this basic process including the addition of up to 25-30 inner layers to produce the more advanced multilayer boards. Multilayer boards with 10-14 layers are much more common.

Printed Circuit Process Sequence

- Bake inner layers, stack up, laminate together
- Rout or shear lamination flash
- Bake - 145°C for 8-16 hr
- Drill and deburr
- Smear removal and etch back
- Electroless copper plate
- Clean and image photoresist
- Electrolytic copper plate
- Tin-lead electrolytic plate
- Remove resist, etch copper
- Fuse tin-lead (reflow)
- Solder mask, nomenclature
- Profile; rout or blank
- Clean
- Test, inspect, ship

6.9.3 The Printed Circuit Cleaning Problem

The chemicals used to produce printed circuit boards cover a wide range from highly alkaline copper plating baths and etchants, to many acidic solutions in plating baths and board conditioning. Most of the chemicals used contain many wetting agents and surfactants. Waste treatment of this chemically intensive manufacturing process is very costly due to labor intensity and the equipment and the plant space required to treat, manifest, and ship away the various effluents.

Hugh volumes of rinse water are used daily to clean PC boards. Water rinsing is used to remove electrolytes and other absorbed chemicals such as solder fluxes and fingerprints that, if left in the laminate, can cause electrical problems.

The industry's main cleaning problem is the removal of electrically conductive absorbed constituents that come from the various processing solutions. These plating and etching solutions contain a spectrum of metal and nonmetalic components that are absorbed into the laminate surface. These absorbed materials must be rinsed out with deionized water in order to conform to one or more electrical specifications (MIL Specs, Bellcore Specs, IP. C specs, etc.). Boards are spot checked for ionic contamination and surface resistivity as a key part of routine QC in all printed circuit facilities.

Part of the PC board specifications are minimum levels of ionic contaminants, which are measured continually on boards at 4-5 strategic process points. The measured ionic contamination is commonly expressed in $\mu g/in^2$ of sodium chloride. The MIL spec is <10.14 $\mu g/in^2$ and the bellcore spec is <6.5 $\mu g/in^2$. The geometries on PC boards are continuing to shrink, making rinsing more critical to meet electrical specifications that are also becoming tighter.

6.9.4 UV Technology

One application of UV laser cleaning in printed circuit board processing is in the cleaning of particulates and organic contaminants from critical layers, for example, copper conductors. The stripping of organic contaminants such as oil or fingerprints results from the ablation of these materials by the absorption of a pulse of deep UV radiation , followed by the reaction of the hot cloud of ablated material with a flowing reactant gas such as oxygen.

A most significant UV cleaning application in PC board production is the replacement of the rinsing step used to remove electrolytes absorbed by the laminate during etching and plating processes. These contaminants must be removed because they result in unacceptable levels of conductivity in the finished board, creating shorts.

UV lasers can replace this rinsing step by shaving off the layer of the laminate containing the contaminants. It has been observed that at high doses UV laser energy will etch the epoxy of a PC board deeply enough to expose the glass fibers.

6.9. PRINTED CIRCUIT BOARDS

Processing at reasonable doses can etch a very thin layer of the laminate in a controllable manner. Since the laminate is an organic material, we have a very clear picture of how this etching occurs. The deep UV radiation is strongly absorbed by the epoxy. A very intense pulse of radiation, absorbed within a period of 10 nsec, and within a depth on the order of a micron, causes this surface material to explosively expand (i.e., ablate).

Depending upon the chemical structure of the laminate, the deep UV radiation may enhance this process by directly breaking chemical bonds in the polymer. With a flow of a reactive gas such as oxygen maintained across the surface where ablation takes place, the hot cloud of ablate material reacts very efficiently, resulting in volatile reaction products which are exhausted with the flowing reactant gas.

In the ablation process, any material absorbed by the laminate will be ablated along with the organic material and form part of the hot cloud. Again depending upon the chemical composition of the contaminants, they may also be reacted in the cloud and exhausted as gases. In any case, they will no longer be present in sufficient densities to recrystallize and redeposit on the board surface. The final result is a PC board laminate surface free of the contaminants which cause shorts in the circuit, at the cost of a microscopic thinning of the board itself, without any detrimental effects on the copper conductors.

It is expected that other application of the technology will be defined. Some of these applications may include the following:

- Roughening of the copper plating base prior to photoresist application to promote adhesion.

- Deoxidizing copper surfaces, for example for improved photresist adhesion or improved contact with the solder layer. Removal of oxide from copper and other metals using UV laser cleaning has been demonstrated, with dilute hydrogen as the reactant gas.

- "Desmearing" or the removal of resin smeared over copper during hold drilling. This is currently done using wet chemical baths, but may be accomplished with the UV laser cleaning process since it requires only the removal of a thin layer of organic material.

6.10 Cleaning Semiconductor Surfaces

Stringent requirements exist for the cleaning of surfaces in advanced semiconductor technology. Most all surfaces become contaminated with airborne particles whose bonding force to the substrate increases as particle size decreases. One of the basic cleaning requirements is to overcome the forces binding results in high energy cleaning mechanisms, especially for particles in the subhalf micron size range.

Cleaning wafers and masks also require that the cleaning process does not deposit a new contaminant of its own. For example, many cleaning solvents contain particles, ions, and trace impurities that can be left behind to form residual monolayers. These new contaminants may be more difficult to remove than the original materials.

Cleaning, whether by laser, chemical, mechanical, ultrasonic, or some combination, must not attack or disturb the semiconductor surface. Many cleaning mediums generate microetching reactions and can attack highly doped oxides or metals.

If high temperatures are used in cleaning, they can result in warpage and physical distortion of the substrate, as well as changing junction depths. Finally, cleaning processes should contain as few steps as possible, as each step adds handling and increased exposure to additional contamination. Ideally, surfaces are processed directly from CVD, metalization, oxidation, or epitaxy without the need for cleaning.

6.10.1 Semiconductor Surface Types

The yield of good parts on a semiconductor manufacturing line is impacted by the cleanliness of the surface. The relationship between defects per unit area and device yield is well documented, as shown in Figure 6.12.

The types of surfaces used in microelectronics are many, but all of them require frequent cleaning before, during, and at the completion of a given process sequence.

Polished silicon wafers, overcoated with a thin layer of silicon dioxide, are the most common surface in semiconductor processing. This is a fairly durable surface and can withstand rigorous mechanical force. However, the polished surface is optically smooth and should not be scratched.

Glass and quartz masks and reticles are another common surface. As photo tools, their cleaning is even more critical. Since defects on masks will be repeated on every wafer printed with that mask, cleaning is required both

6.10. CLEANING SEMICONDUCTOR SURFACES

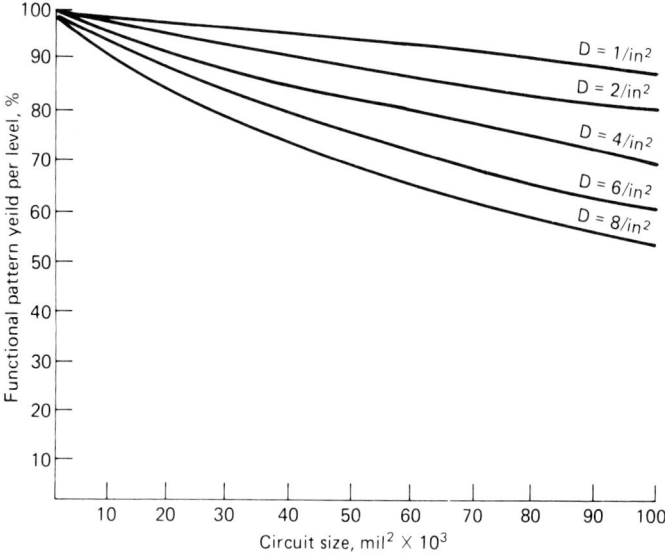

Figure 6.12: Defect Density vs Yield in Semiconductor Manufacturing

before patterning the chrome layer and after patterning. In both instances, the surface is delicate, mirror-like, and easily contaminated by thin liquid films that can optically distort the images transmitted through the mask.

Pellicles are extremely delicate membranes that catch particulates before they land on a mask or reticle. They are mounted above the surface of a mask or reticle at a distance that keeps the particles out of focus in the optical path. Pellicles are an example of materials that require noncontact cleaning. Pellicles, like masks, will cause printable defects if their surface is sufficiently scratched or damaged.

Thin film circuits can have very high density circuitry and may be difficult to clean with conventional techniques. Thin film circuits have a variety of metals and nonmetals on their surfaces; laser cleaning, unlike chemicals or gases, does not differentiate if several metals and/or nonmetals are in a surface. Since the metals and ceramics used in this technology have high ablation thresholds, they can be cleaned in a single step, ablating debris off a variety of different materials from the surface.

A completed device, before or after packaging, may require cleaning. Noncontact laser ablation of preformed, delicate structures is ideally suited in these instances. High aspect-ratio, pre-etched devices likewise require noncontact cleaning.

Finally, there are substrates that cannot be subjected to any wet chemical or mechanical cleaning method. Certain plastics, as well as devices used in ultra-clean vacuum environments, can be cleaned with UV laser irradiation.

The main condition necessary to allow efficient use of UV laser surface cleaning is differential ablation threshold, i.e., sufficient difference between the ablation threshold of the substrate and the material being ablated.

Most contaminants are small enough to be transported away by the acoustic excitation and evaporation of the water or alcohol fluid that acts as an interface. Thus, the UV photons are not necessarily used to dissociate the contaminant into molecular or macromolecular fragments, as it does with the fluid. The ablation threshold of the interface fluid is then the critical parameter in optimizing the process.

Contaminants, organic and inorganic, can be removed with relatively low fluence UV energy. Also, the ablation thresholds of the substrate materials are considerably higher, thereby allowing a considerable margin for overexposure in cleaning without damaging the substrate.

6.10.2 Cleaning Guidelines

The guideline for cleaning is to avoid a cleaning step and to use a freshly deposited or freshly etched surface if possible; cleaning with alcohol rinses or spin/flood rinses with resist thinner (before resist coating) is used to "scrub" off lightly bound airborne particulates. Following a cleaning step, wafers or masks and reticles can be stored in a nitrogen-purged dessicator or similar environment. Even slight amounts of contamination can disrupt the performance of a high density IC device. For example, a monolayer of sodium contamination equal to 1×10^{-4} ions inverts the surface of a 1-Ohm/cm silicon device. Since MOS devices rely mainly on surface state operation, they are especially sensitive to contamination.

Wet chemistry cleaning solutions leave behind residues and "dry" cleaning is often preferred. Dry oxidation is such a technique. This can be done in a 900-1200° oxidation tube (1), an ideal reducing environment. Care needs to be taken that the temperature does not change junction depths. Oxygen plasmas are also used for resist removal.

Ultrasonic and megasonic cleaning is another "noncontact" cleaning approach. Sonic wave energy is effective in removing some particles, but is restricted because most solvents leave residues. Isopropyl alcohol is an ex-

6.11. LASER ABLATIVE CLEANING

ception. Most all of the techniques mentioned earlier are only marginally effective for the removal of tightly bound submicron particles.

6.11 Laser Ablative Cleaning

Photoablation of organic and other debris has been demonstrated with 193-, 248-, and 308-nm excimer lasers (2,3). Excimer lasers provide intense UV energy that is strongly absorbed at the surface of most semiconductor films. This absorption, as energy coupled into any of the surface contaminants (including metal particles) typically found on wafer, mask, reticle, or pellicle surfaces, is sufficiently intense to dissociate the absorbing material by photoablative decomposition.

Initial experiments with excimer laser cleaning by photoablation showed that four laser pulses at a low fluence (0.35 J/cm^2) irradiation at 248 nm were sufficient to remove aluminum particles from a contaminated mask surface (2). In this section we will discuss the various approaches to the excimer laser cleaning of semiconductor surfaces and will review the results of experimental work performed in several applications.

6.11.1 UV Laser Removal Mechanisms

The basic mechanism for removal of particulates from semiconductor surfaces is intense absorption of UV photons by the particles, followed by expansion of the particle. The rapid acceleration of the particle (10^{10} cm/sec^2) is the result of atomic vibrational and rotational excitation caused by the intense UV absorption. The heat produced by the pulsed laser exposure drives the reaction. Expansion is followed by acceleration and expulsion of the particle away from the surface. The UV absorption produces sufficient energy within the particle, in the form of megasonic vibrations, to overcome van der Waals forces adhering the particle to the surface.

Many types of particles can be ejected, normal to the surface on which they reside, by laser irradiation. This reaction occurs without melting or damage to the underlying layer. After ejection, the particles slow down rapidly within about 10 mm of the surface. The fluence required to produce this reaction is dependent on the type of particle being irradiated. For example, low density organic particles are removed with less than 0.3 J at 248 nm; metal particles, ceramics, and other inorganics require higher fluence. The fluence must also exceed the van der Waals force, which is typically seven to eight times the gravitational force (4).

The size of the particle, particle population density, and surface type also determines the ablation threshold. Small particles, with greater surface contact and higher van der Waal adhesion energy, have higher ablation thresholds than larger particles of the same type under identical conditions.

The removal mechanism rate occurs also as a function of laser rep-rate and pulse length. Shorter pulses are more effective for particle removal and provide more latitude for overexposure without causing substrate damage. For the same reason, low rep-rates (<100 Hz) are more efficient, allowing for better thermal recovery and equilibrium, and reducing the amount of incident beam interference with plumes of effected debris or ablation by-products. The average pulse length for krypton fluoride excimer lasers is 10-20 nsec; this is also the typical pulse length of 193-nm excimer lasers used for the same application.

Experiments with infrared laser cleaning have shown that only slight cleaning occurs. Energy from infrared lasers is absorbed at the surface of the particle or films rather than at the silicon (or other substrate) interface. The pulse length of infrared lasers is short (Q-switched Er:YAG at 2.94 μm), but the effects do not cause the intense and rapid absorption as occurs at the short ultraviolet wavelengths. Instead, heating effects occur, and, in the presence of water or other fluids, surface boiling with little to no ejection force from the substrate interface. Increasing the fluence of an infrared laser only increases the amount of melting.

6.11.2 Dry Cleaning vs Wet Cleaning

Dry particle removal relies solely on the interaction of the UV photons with the particles or contamination being irradiated by the laser. The addition of other liquids, such as water, may serve to increase the cleaning efficiency by increasing the number of ejection reactions. UV laser energy can produce an explosive boiling effect on water films. It is believed that capillary water condensation occurs between the surface below the water film; the wafer or mask containing the contamination is covered with a thin film of water or other fluid.

The presence of a liquid film, and the absorption of UV photons both in the substrate and in the film, causes a maximum ejection force away from the substrate, carrying away particulates simultaneously with the vaporized water film. An additional benefit of the wet cleaning method is that the substrate surface is cleaned along with the removal of surface-adhered particle contamination. Figure 6.13 shows the results of samples before (left photo) and after (right photo) UV laser irradiation.

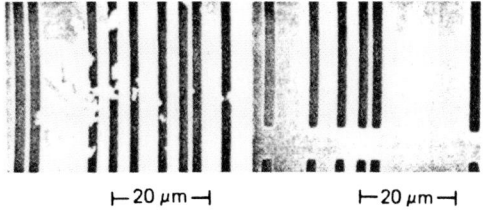

Figure 6.13: UV/Steam Cleaning of Alumina

The samples referred to in Figure 6.12 show how the Al_2O_3 particles can be removed without damaging the 2-μm-thick membrane. The particle size is about 0.35 μm, removed with a 248-nm excimer laser. The number of pulses required to remove these particles varies from 1 to 5, with a pulse length of 16 nsec. The beam was normal to the surface during exposure, with a spot size of several millimeters (2).

6.12 Debris Removal

Laser cleaning, to be completely effective, must provide a means for debris removal. Scavenging nozzles used with photoablation systems typically produce both a positive and a negative air or inert gas stream to direct the plume of debris away from the substrate. Ablative debris plumes tend to have a characteristic shape and a predictable pattern of velocity.

In a typical manufacturing process, wafers are cleaned at several stages: preoxidation (after wafer polishing), prephotoresist coatings, and post-etching or pre-ion-implantation. The types of contaminants found at each stage will change according to the environment the wafer is exposed to. Reworked wafers, where resist is stripped and cleaned, are another common place for cleaning.

Particles interfere with the growth of oxide or other layers, and also can disturb the adhesion of resist layers.

Small particles are tightly bound to the wafer surface by van der Waals forces. Various types of interatomic forces hold particles to semiconductor surfaces, including hydrogen and covalent bonding forces. Figure 6.14 shows potential energy curves for these forces.

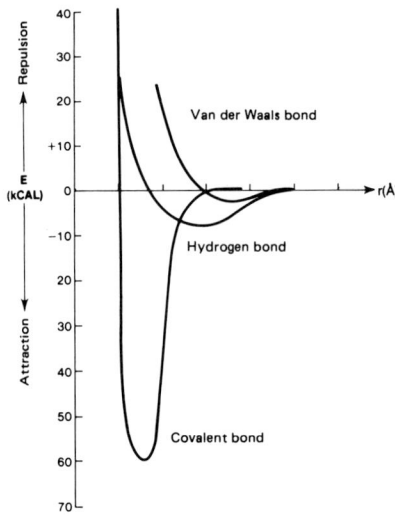

Figure 6.14: Potential Energy Curves for Different Types of Interatomic Forces

The bonding forces that hold micron and submicron particles on microelectronic surfaces need to be overcome in cleaning without causing damage to the surface. High temperatures, harsh acids, and high intensity water streams all have the potential of damaging preformed structures on the IC, mask, or similar semiconductor surface. Solvents and cleaning solutions, including ionic and nonionic surfactant-based detergents, may leave residual residues or monolayers. These films are extremely thin (200-1000 Å), are difficult to detect, and may be even more difficult to remove.

6.12.1 UV Laser Cleaning Equipment

The basic requirements for UV laser surface cleaning include a laser and appropriate beam delivery optics, a three-motion substrate stage system, an apparatus for the collection of ablative debris, and a viewing system for introducing the liquid or gas to be associated with the laser in the cleaning process. The apparatus shown in Figure 6.15 is typical of a system containing the basic functionality required for cleaning a wafer, mask, or other flat surface.

In the system shown, a 248-nm excimer laser is used, but a 193-nm laser

6.12. DEBRIS REMOVAL 203

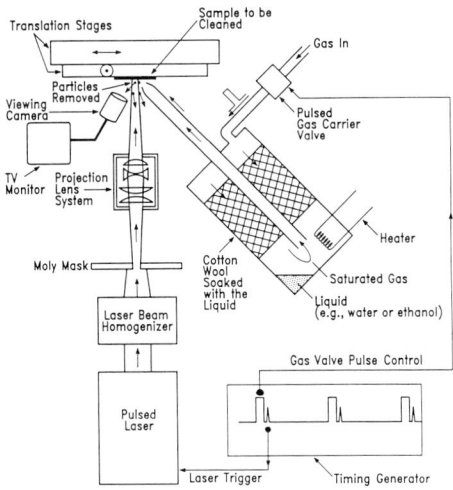

Figure 6.15: UV Laser/Steam Cleaning System Schematic

or any similar pulsed or CW deep-UV radiation source could be used, including a 157-nm fluorine laser. The beam is processed through an optical homogenizer and is sent through a molybdenum mask which serves as an aperture to mask the beam and form an aerial image if necessary. A projection lens then images the beam onto the substrate. A video camera records the sample movement and exposure reaction, and the XYZ translation stage manipulates the sample under the beam.

The system shown employs a pulsed liquid film deposition system to coat the sample just before laser irradiation. A pulsed gas valve injects nitrogen into a cylinder that supplies heated ethanol, isopropanol, or water vapor onto the substrate. The liquid is dispensed according to a pulsed timing unit which dispenses the fluid approximately 0.1 sec before triggering the laser. The timing circuit that triggers the laser and injects the fluid can repeat the cleaning cycle several times.

The various types of contamination encountered in semiconductor processing may require different laser wavelengths at different fluence levels from those cited in the above data. The spectral absorption profile of the material to be removed may be important, especially if a continuous film must be removed.

6.12.2 UV Laser and Gas Interaction

UV photoreactive cleaning is a technology that uses high energy photons, focused into a blade at the substrate (10) to excite specific gas chemistries for the purpose of cleaning, stripping or other surface conditioning. UV wavelengths of 266, 248, and 193 nm have been proven effective in the removal of a wide variety of organic contaminants, using simple "green" or non-polluting gases like oxygen, nitrogen and even filtered air: The use of UV/gas cleaning provides potential replacement for conventional wet cleaning which involves not only highly corrosive chemicals and solvents, but also requires large volumes of rinse water. The UV/gas technology may, in certain cases, produce non-volatile byproducts, but these can be trapped in a simple carton exhaust filter.

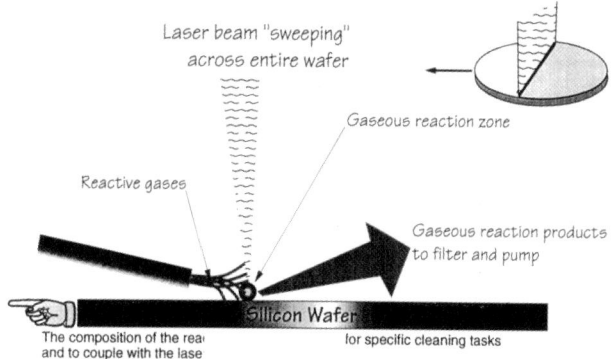

Figure 6.16: Schematic of UV/Gas Photoreactive Cleaning Concept

UV photoreactive dry cleaning can be implemented by focussing a UV laser beam with a 3-element optical system into a thin blade of light, and injecting a stream of gas(es) in the path of the incident light. A UV laser photoreactive cleaning system schematic is shown in Figure 6.17. The system uses a small gas reaction chamber into which UV light is injected through a quartz window. The substrates may be heated with a temperature controlled chuck, and gas mixture may also be heated before being set into the "gas reaction zone" where they interact with high energy UV photons to form a plasma.

6.12. DEBRIS REMOVAL

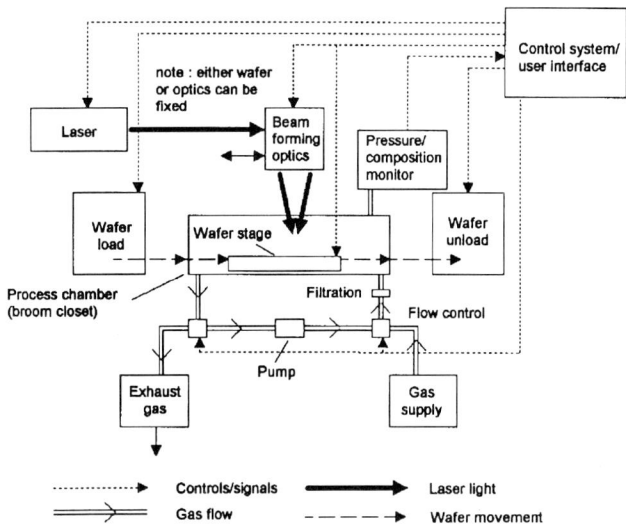

Figure 6.17: Schematic of a Photoreactive UV/Gas Cleaning System

References

1. Elliott, D.J., "Integrated Circuit Fabrication Technology," Second Edition, McGraw Hill, Chapter 5, page 143-178, 1993.

2. Zapka, W. Ziemlich, W., Tam, A.C., "Efficient Pulse Laser Removal of 0.2um Sized Particles from a Solid Surface," Applied Physics Letters, 58, (20), May 20, 1991.

3. Elliott, D.J., Piwczyk, B.P., "Material Processing with Excimer Lasers," MRS Boston Technical Program, Dec. 1986.

4. Bowling, R.A., "Particles on Surfaces," ed K.L. Mittal, Plenum, NY, 1988, Volume 1, page 129.

5. Imen, K., Lee, S.J., Allen, S.D., Cleo, 1990.

6. Harte, K., Image Micro Systems, Private Communication.

7. "Laser Cleaning Technologies for Removal of Surface Particulates," Journal of Applied Physics, Vol. 71, No. 7, April 1992.

8. Data and photo courtesy of Kathy Logan, Tencor Instruments.

9. Ruzylo, R., Novak, R.E. (eds.), "Proceedings of the First, Second and Third International Cleaning Symposium on Cleaning Technology in Semiconductor Manufacturing," Elec. Chem. Soc., 1990-94.

10. UVTech Systems, Inc., Wayland, MA; Internal Publication, July 1995.

Glossary of Terms

APM A mixture of ammonium hydroxide, hydrogen peroxide, and deionized water for light organic contamination removal.

Ashing The use of oxygen (or other) plasma in an RF field to strip organic or metal/semiconductor filter films.

Brownian Motion The suspended movement of microscopic particles caused by the force of their surrounding medium (liquid or gas). For example, submicron particles whose individual mass is smaller than the mass of the air molecules surrounding them will be suspended and will move as a function of the movement of the surrounding gas molecules. This phenomenon was discovered by Robert Brown in 1827.

CMP Chemical, mechanical polishing; a planarizing method that leaves a dirty particle-ridden residue.

DHF Dilute hydrofluoric acid.

IPA Isopropyl alcohol, used to promote drying of IC surfaces as an additive to aqueous cleaning mixtures.

Organometallic An amalgam of metallic and organic deposits from etching or deposition processes.

SC-1, SC-2 Standard clean 1 and 2, hydrogen peroxide-based wet cleans developed at RCA; also called "RCA standard cleans."

SPM A mixture of sulfuric acid, hydrogen peroxide, and deionized water for IC removal of heavy organic contamination.

van der Waals Force An attractive force between two atoms or non-polar molecules, caused by interacting dipole moments. van der Waals forces help keep particulates on semiconductor surfaces.

Reading Bibliography

1. Frerichs, H., Tricker, J., Wesner, D.A., Kreutz, E.W., "Laser-Induced Surface Modification and Metallization of Polymers," Applied Surface Science, Vol. 86, p. 405-10, Feb. 1995.

2. Laude, L.D., Kolev, K., Brunel, M., Deleter, P., "Surface Properties of Excimer-Laser-Irradiated Sintered Alumina," Applied Surface Science, Vol. 86, p. 368-81, Feb. 1995.

3. Viville, P., Thoelen, O. Beauvois, S., Lazzaroni, R., Lambin, G., Bredas, J.L., Kolev, K., Laude, L., "Thin Films of Polymer Blends: Surface Treatment and Theoretical Modeling," Applied Surface Science, Vol. 86, p. 411-16, Feb. 1995.

4. Kuzmichov, A.V., "Excimer Laser-Assisted Etching of Silicon in Chlorine: Adsorption and Desorption," Applied Surface Science, Vol. 86, p. 559-63, Feb. 1995.

5. Henari, F., Blau, W., "Excimer Laser Surface Treatment of Metals for Improved Adhesion," Applied Optics, Vol. 34, No. 3, p. 581-4, Jan. 1995.

6. Kumuduni, W.K.A., Nakata, Y., Sasaki, Y., Okada, T., Maeda, M., Kisu, T., Takeo, M., Enpuku, K., "Effect of Cumulative Ablation on the Ejection of Particulates and Molecular Species from $YBa_2Cu_7 - x$ Targets," Journal of Applied Physics, Vol. 77, No. 11, p. 5961-6, June 1995.

7. Watanabe, J.K., Gibson, U.J., "Excimer Laser Cleaning and Processing of Si (100) Substrates in Ultrahigh Vacuum and Reactive Gases," Journal of Vacuum Science and Technology A (Vacuum, Surfaces, and Films), Vol. 10, No. 4, Part 1, p. 823-8, July-Aug. 1992.

8. Venu Menon et al (Sematech) "Technology Trends in Silicon Wafer Cleaning," Vol. 15 of MRS, Spring 1993.

9. Ohmi, T., "Tohoku University Research on Wafer Surface Cleaning," Institute of Environmental Science p. 255, 1993.

10. Iscoff, R., "Wafer cleaning: wet methods still lead the pack," Semiconductor Interntional, p. 58, July 1993.

11. Kern, W., "The evolution of silicon wafer cleaning technology," Journal of Electrochemical Society, Vol. 137, p. 1887, June 1990.

12. Singer, P.H., "Trends in wafer cleaning," Semiconductor International, p. 36, December 1992.

13. Kern, W., Ed., "Handbook of Semiconductor Wafer Cleaning Technology: Science, Technology Applications," Noyes Publications, 1993.

14. Ohmi, T.,Ed., "Ultraclean Technology Handbook," Vol. 1: Ultrapure Wafer, Marcel Dekker, Inc., NY, 1993.

15. "Semiconductor technology workshop: working group reports," Semiconductor Industry Association, San Jose, CA 1993.

Chapter 7
Annealing and Planarizing

7.1 Introduction

The reduction of circuit geometries in VLSI device fabrication led to the introduction of ion implantation as a space-conserving method of doping the wafers with controlled amounts of impurities. The high energy ion beams caused damage to the crystalline surface which required repairing. Originally, pulsed ruby lasers were used to anneal out the ion implantation-induced damage to the silicon crystal lattice. This was reported by scientists in Kazan, Russia sometime before 1976, when it was first discussed in the United States.

Ion implantation is a process where the imaged and oxide-etched wafer surface is bombarded with a high energy beam of dopant ions such as boron or arsenic. The energy of the beam is controlled to provide sufficient kinetic energy to drive these impurities, and impurity ions can be driven into the silicon crystal in controlled quantities with uniform distribution and to very specific depths. This ability in turn allows IC circuit designers to reduce circuit dimensions for higher performance devices (shallower junctions, higher speed, smaller line widths).

Figure 7.1 shows the difference in channel gate length between conventional furnace annealing (A), where thermal diffusion spreads the dopant ions laterally in the silicon, and laser annealing (B), where dopant ions remain held within the lateral dimension defined by the ion implantation step. Ion implantation has displaced furnace diffusion as the production method for doping semiconductor devices.

The key problem caused by ion implantation is that the kinetic energy

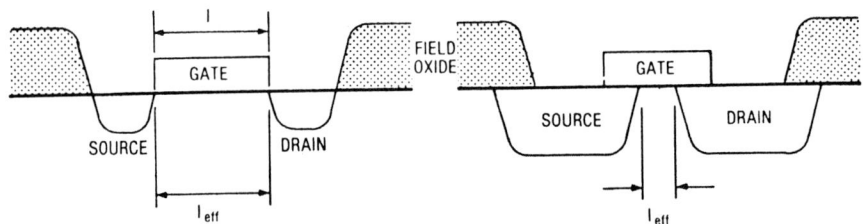

Figure 7.1: Channel Gate Length: Furnace Anneal vs Laser Anneal

or momentum of incoming ions rearranges the silicon crystal atoms, disordering them to an essentially amorphous state. This rearrangement of the crystallinity of the silicon undesirably changes the electrical properties of the device.

Furnace annealing is used to reheat the wafer, thereby melting the silicon and allowing the crystal lattice to reform its natural crystalline order. Furnace annealing is not practical for advanced IC devices since the IC dimensions are so small that thermal distortions throughout the wafer cause dimensional run out and loss of critical dimensions. Laser annealing heats only a thin layer of the surface of the wafer, preventing thermal conduction into the bulk of the silicon. Laser annealing does not induce nonflatness (warp, bow) into the wafers.

Another related use for AUV lasers (aside from reordering the lattice after damage caused by ion implantation) which is scanned across the entire wafer surface is dopant activation. Dopant activation requires sufficient thermal energy to allow boron or arsenic atoms to take substitutional positions in the silicon lattice. This requires slightly more energy than required for recrystallization and causes some distortion of the silicon lattice as silicon atoms must "move over" to make room for the impurity atoms, which may be larger or smaller than the atoms they are displacing.

7.2 Annealing Parameters

Laser annealing parameters are determined by several factors. Unlike furnace annealing where the amount of energy absorbed is independent of depth, the laser light energy used is absorbed in a thin layer on the surface of the wafer. The major determinate of energy intensity in this top layer is the

7.2. ANNEALING PARAMETERS

optical absorption coefficient of the wafer material. The actual energy flux or intensity that reaches a given depth (x) is expressed as follows:

$$I(x) = I_o(l - R)e^{-\alpha x} \tag{7.1}$$

where

I_o = Incident Light Intensity
R = Surface Reflectivity
α = Absorption Coefficient

Differentiation of the above equation permits calculation of the energy deposited at x depth. The differentiation equation is:

$$\mid dI(x) \mid = I_o(1 - R)\alpha \epsilon^{ax} dx$$

This equation shows that most of the energy will be deposited within a layer of thickness $\frac{1}{\alpha}$, that the majority of the energy will be greatest at the surface ($x = o$), and that energy intensity will decrease exponentially with depth. Finally, the above equation assumes a homogeneous material, which in practice in not always the case.

The absorption coefficient is thus a big parameter in optimizing the annealing process time or duration of laser expense on a given site on the wafer surface. The laser wavelength and silicon crystal orientation, including the specific absorption of the silicon, determine the absorption coefficient. Wavelength dependence for absorption is strong in semiconductors, changing rapidly with energy equal to the band edge of the specific semiconductor. The equation to express this relationship is:

$$\lambda_c = 1.245/E_g \tag{7.2}$$

where

λ_c = wavelength (μm)
E_g = Band Gap in Electron Volts

Laser-transparent semiconductors (low absorption coefficient) are those whose L exceeds L1, i.e., a small (< 1 cm^{-1}) absorption coefficient. Lasers, such as Nd:Yag, argon, and ruby, emit at wavelengths shorter than L and result in absorption coefficients on the order of 10^{-3} to 10^{-4} cm on the wafer surface. In the case of crystalline silicon, the wavelength at which the absorption coefficient changes is 1.1 nm. Excimer lasers, with much shorter wavelengths, are capable of providing even more rapid and efficient annealing in even shallower layers of the wafer surface, reducing further the thermal effects seen with ruby and Nd:Yag lasers.

The crystal orientation of the silicon radically changes the absorption coefficient. Amorphous silicon, for example, has an absorption coefficient of 5000 cm^{-1} at 0.69 μm (ruby laser), while under the same irradiation conditions, single crystal silicon has a value of 500 cm^{-1}. The speed of the reaction is so great with laser annealing, compared to furnace annealing, that very little energy is conducted into the bulk of the wafer. This efficiency contributes favorably to the economics of laser annealing. As energy costs in semiconductor processing became a larger fraction of the total process cost, laser annealing at even shorter and more thermally efficient wavelengths is practical. For example, a 266-nm quadruple of a Nd:YAG laser with a focused blade could be used.

Finally, semiconductor layers that require annealing, such as thermal grown field oxides, are constantly being reduced in thickness as devices become more three dimensional and overall density increases. Rapid laser annealing, with a high degree of control over the Z-dimensional thermal profile, is a much needed enhancement in VLSI device fabrication.

A plot of normalized energy density absorbed in silicon versus depth in micrometers is shown in Figure 7.2.

Note that the amount of the energy absorbed in the case of electron beam annealing is driven strongly by the amount of kinetic energy in the beam. While the curve for laser or optical absorption shows a much less increase in depth for a given change in beam energy, sufficient heating occurs at the required depths to make lasers useful for most applications in VLSI.

Laser annealing efficiency is also dependent on the reflectivity of the surface while electron beam annealing is sensitive to electrical effects. Laser annealing has the advantage of being a nonvacuum, atmospheric process since the light (unlike the electron beam) is not noticeably deflected, scattered, or absorbed by air molecules.

The sensitivity to optical reflections in laser annealing is reduced with shorter wavelength lasers as the annealing is effected closer to the surface and as less energy reaches the deeper structures (buried emitter) with their varying reflectivities.

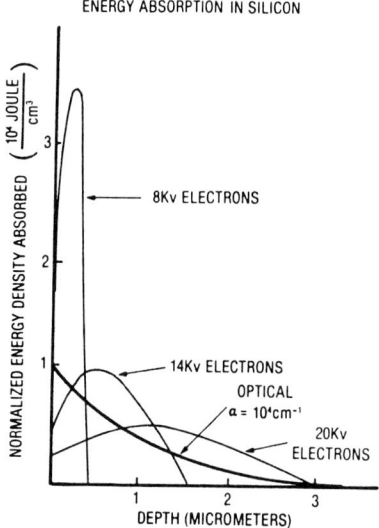

Figure 7.2: Normalized Energy Density vs Depth in Laser Annealing

7.3 Laser versus Electron Beam

Electron beam annealing offers two key advantages not provided by laser annealing. A beam of electrons is not affected by differences in surface reflectivity and the deposited energy does not decrease exponentially with depth, as it does with a laser. The method of energy transfer in electron beam annealing is through scattering reactions, during which the incident electrons transfer a portion of their energy.

A beam of photons, by contrast, transfers its total energy to electrons as a single quantum or does not transfer at all. As an electron beam penetrates a silicon layer, it releases energy and reaches maximum absorption typically within 30% of the maximum depth the electrons reach. This characteristic allows a process operation to preselect an optimum depth by varying the beam energy at the outset.

Lasers, on the other hand, provide the maximum energy at the surface, with energy levels falling off exponentially with depth. This has more overall advantages in semiconductor processing since only in isolated cases is maximum energy needed somewhere below the surface. For example, group III semiconductors are more difficult to anneal with maximum energy at the surface.

7.4 Pulsed versus Continuous Lasers

Pulsed and continuous laser annealings are used in semiconductor processing. The mechanism of pulsed annealing is believed to be a moving liquid: a solid interface. The laser melts the surface first, and the melt depth grows until it reaches the single crystal silicon; at which point the regrowth back towards the surface begins as the wafer cools. This moving melt interface, with the single crystal layer acting as the "seed," proceeds at a velocity of 1 m/sec.

The melt moves first from the surface down to the single crystal boundary and back to the surface as the newly reordered amorphous silicon cools. The length of time the silicon is in a molten state is critical, as lateral diffusion of dopant impurities will occur. Pulsed lasers at relatively long wavelengths (0.7-1.06 nm) achieve a complete anneal in 0.05-0.1 nsec. This is relatively a long time in terms of the criticality of lateral diffusion. Shorter wavelength excimer lasers, with very short pulse lifetimes (nsec-psec), may achieve the anneal in much shorter time and with much greater efficiency of cost, throughput, and equipment.

Pulsed annealing is typically performed with Nd:YAG (neodynium-doped yttrium-aluminum garnet-rod) and ruby (chromium-doped aluminum oxide rod) lasers.

Continuous annealing is performed with argon (glass/ceramic vessel filled with gas) lasers. In continuous annealing, a solid-solid interface is the mechanism for amorphous silicon regrowth or dopant activation. The advantage of restricting lateral dopant redistribution is important, as continuous laser annealing is an epitaxial film growth process. The photons are absorbed first at the crystalline/amorphous (or damage) interface and crystal growth proceeds toward the wafer surface. Thermal energy, but not melting, provides the means for atomic rearrangement.

Continuous annealing with solid phase growth is comparatively slow, approximately 4 orders of magnitude longer than the 0.1-nsec interaction time for a Nd:YAG laser. This adds considerable cost for production, as many more systems need to be purchased to match the throughput reached with pulsed annealers. Another parameter restricting throughput is beam spot diameter, typically over 2 orders of magnitude smaller with argon lasers (0.05 mm versus 1-10 mm for pulsed lasers).

Excimer lasers offer larger beam sizes for equivalent energy densities than longer wavelength lasers, increasing throughput still greater. The use of large diameter silicon wafers and overall increases in IC production annually make throughput an important economic issue in laser annealing.

7.5 Annealing Process Control

The main functions for lasers are annealing out ion implantation damage and activating dopant ions in the lattice. Epitaxial growth and liquid melting both achieve high quality crystalline regrowth, with the lack of dependence on single crystal quality and orientation favoring pulsed annealing. Pulsed and continuous wave laser processing requires optimization of the main input parameters. These include spot size, energy density in the beam and in the film (silicon, gallium arsenide, etc.), laser wavelength selection to effectively best couple energy into the wafer, and the optimal exposure time.

Quality control before (optimization of parameters), during (in-site metrology), and after (changes in refractive index, extinction coefficient) is necessary to key the final IC device electrical characteristics within specifications.

Nondestructive measurements of optical reflectivity before, during, and after annealing are used to determine the extent of the anneal process time and quality of the process. Before annealing, measurements of the refractive index and extinction coefficient of ion-implanted amorphous silicon are taken. The typical values for amorphous silicon and crystalline silicon are shown below.

	Refractive Index	Extraction Coefficient
Amorphous Si	4.5	0.7
Crystalline Si	3.8	0.1

Amorphous silicon thus reflects more light than crystalline silicon and is therefore visibly lighter.

Since even small changes in the crystallinity of the silicon (or other semiconductor material) will result in easily measured changes in the optical extinction coefficient and refractive index, measuring these parameters provides a high degree of quality control. Constructive and destructive light interference occurs as light is reflected back from the two layers (surface, amorphous/crystalline boundary) of the wafer, with light intensity increasing as the crystalline boundary approaches the wafer surface. This occurs since the amorphous layer is very highly absorbant. As it becomes thinner during the anneal, it is closer to the surface and therefore more visible.

Surface reflectivity is measured with a helium-neon laser, and plots of the light maxima/minima are made. This optical metrology technique provides good characterization of the depth of the implantation before and after annealing.

7.5.1 Alpha Partick Backscattering Analysis

A nondestructive technique for measuring the variation in crystalline quality is Rutherford backscattering analysis. This test involves exposing the crystal to a stream of lightweight atomic nuclei at an angle normal to the crystal axes.

The incoming nuclei will undergo collisions between atoms and nuclei in the sample. In a layer of high quality crystalline silicon, the nuclei will travel along the channels in the atomic lattice, with relatively few collisions. In amorphous silicon, there are few such channels, and many collisions will occur, with energy being backscattered. Typically, the lightweight nuclei reflected will be alpha particles of helium ions and protons of hydrogen ions. The analysis of the reflections provides information about the character and population of imperfections in the semiconductor layer.

Energetic nuclei penetrating amorphous silicon will still move through the material, since even disordered crystal lattices have relatively wide open spaces (on a nuclear scale). Coulomb interactions will cause the incoming positively charged nuclei to lose energy, and their energy spectrum will reflect this loss of energy. Thus, Rutherford backscattering analysis provides information about crystalline disorder or quality, depth and thickness of the imperfections, and distribution of these damaged or newly annealed sites in semiconductor films.

Figure 7.3 shows energy distribution profiles of a sample processed before and after ion implantation, and after laser annealing.

The points of the curves measure the backscattered kinetic energies of alpha particles. The incident beam was directed with a <110> crystal. The five curves (labeled on the figure) are, in the process sequence in which they occur, as follows:

- Virgin Crystal

- 35-kV Boron Ion Implant

- Ruby Laser Annealed (1.75 J/cm^2)

- Furnace Annealed (900° C)

- Random

7.5. ANNEALING PROCESS CONTROL

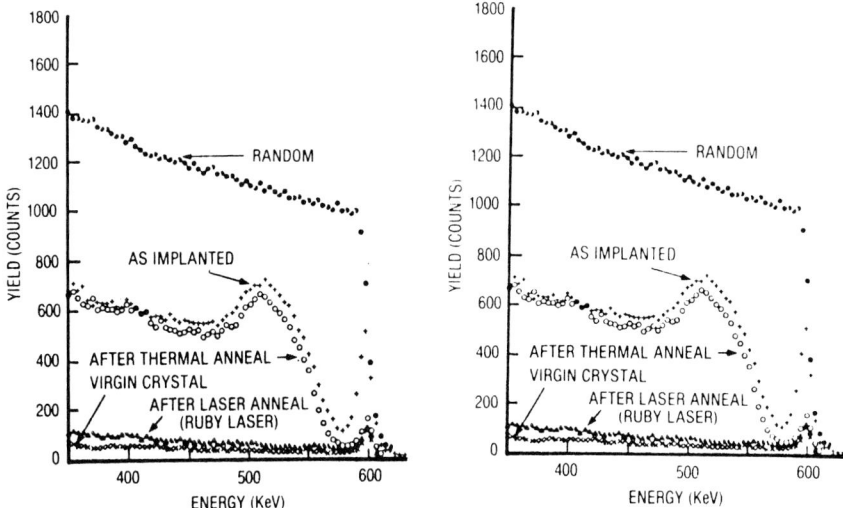

Figure 7.3: Energy Distribution Profiles before Ion Implantation and after Laser Annealing

Note that laser annealing restores the silicon crystallinity to nearly the same level as the virgin silicon, now including the added dopant ions.

7.5.2 Doping and Dopant Activation

In dopant activation, UV lasers may also be used for doping and for dopant activation. Figure 7.4 shows a schematic of a wafer being doped by a UV laser. The dopant specie is absorbed onto the silicon surface and is driven in by one of three possible steps: activation, diffusion, or regrowth of an oxide which provides the necessary thermal energy for diffusion. The doping step is initiated by a 193-nm excimer laser pulse.

The amount of dopant activation is determined by measuring the sheet resistance of the device. Sheet resistance is the resistance between two parallel edges of a square sheet of material of the conductive layer. Sheet resistance is expressed in ohms per square, according to:

$$R_s = \rho/d$$

where

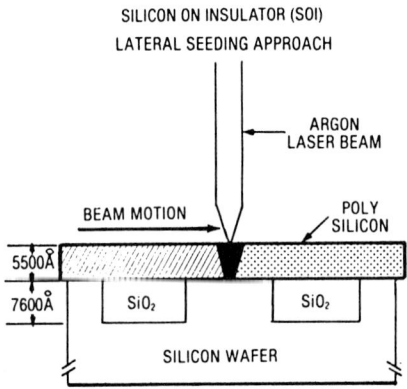

Figure 7.4: Laser Doping Schematic

R_s = Sheet Resistance of a Homogeneous Layer
ρ = Electrode Resistivity
d = Layer Thickness

Measurements of sheet resistance in furnace-annealed versus laser-annealed samples showed equivalent values after processing. The major difference is that furnace-annealed samples undergo lateral dopant redistribution, while laser dopant-activated specimens retain the same dopant distribution as existed after implantation.

Resistivity of a given semiconductor material is a function of charge carrier concentration and carrier mobility, expressed as follows:

$$R_s = \rho/d = \frac{1}{neud}$$

n = Carrier Concentration
e = Electronic Charge on Carriers
u = Carrier Mobility

This assumes that "n" and "u" are constant in the film layer and that the layer is uniform and thick.

Dopant distribution in silicon layers can be measured by SIMS (secondary ion mass spectroscopy). This analytical technique is limited to measurement

7.5. ANNEALING PROCESS CONTROL

close to the wafer surface, but will reach deep enough to monitor dopant distribution and profile. A typical SIMS plot, comparing laser- and furnace-annealed samples, is shown in Figure 7.5.

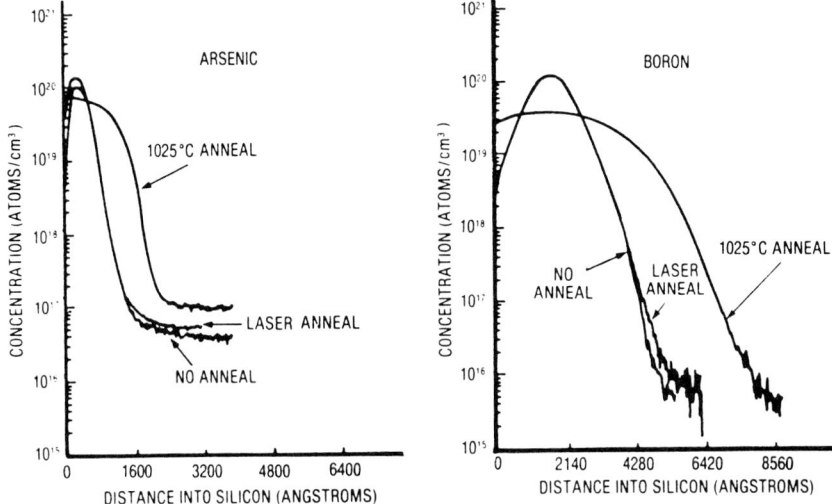

Figure 7.5: SIMS Profiles of Laser- and Furnace-Annealed Samples

The spatial distribution of the impurity ions using laser annealing almost matches the profile before the annealed step. At the tail of the curve, there is a slightly earlier fall-off of concentration in the laser-annealed sample compared to the "no anneal" sample. For an equivalent depth, furnace annealing leaves a slightly higher concentration at the 1025 °C temperature; only in the top 1-2000 Å does laser annealing exceed the dopant concentration of the furnace anneal.

7.5.3 Summary

The use of lasers for annealing out crystal damage caused by ion implantation has been driven largely by the overall need to reduce the amount of high temperature processing by either eliminating high temperature steps or reducing the maximum temperatures used. Thermal distortion of the wafer caused by high temperature process steps produces dimensional "run out" (wafer warping) and will reduce chip yield at the probe step. The specifications for CD (critical dimension) control are continually tightened to meet a total overlay (all mask-levels summed) budget. Wafer diameter is also continually increasing, a factor that increases the need to reduce high temperature processing where possible.

Lasers keep the primary thermal energy within the area of functional need, i.e., near the surface, and are not only more thermally efficient as a result, but avoid warping the wafer by not heating the bulk, as is the case in furnace processing.

Photons are energy sources for heating, and reaction initiation has become useful as the layer thickness is reduced in VLSI fabrication. For reasons of contamination control, light-initiated or light-driven reactions are generally cleaner than furnace or similar enclosed environments that involve increased mechanical handling and movement of parts. Matching the absorption coefficients of semiconductor films as closely as possible to the wavelength of the laser provides a pathway to achieve a high level of depth control and reaction rate, both critical parameters in volume production fabrication of IC devices.

7.6 Laser Planarization

The ability to planarize a surface has useful application in many areas of technology. Localized surface melting that is performed with a UV laser source has been used in laser annealing of semiconductor films, recrystalization of wafer surfaces, stress relieving of metal layers, and sealing of low density materials. In this section, we will focus on the use of lasers for planarizing the surface of IC devices. UV and longer wavelength lasers will be discussed and results presented for this application. The same or similar equipment and optics used for planarization of wafer surfaces may be applied to other surface melting applications.

In advanced IC fabrication, multilevel interconnection layers are used to achieve high levels of chip density without increasing chip size. The various metal layers (aluminum, aluminum alloys, silicides) are interconnected by a large number of submicrometer-sized via holes. The via connections between metallization layers are high aspect ratio structures that are difficult to metallize by standard deposition methods. During metal deposition, poor metal coverage caused by shadowing from the via walls can lead to device reliability problems.

The use of multilevel interconnections and the increasing number of individual layers used to fabricate an advanced IC create steep topographic

7.7. IC TOPOGRAPHY PROBLEMS

features. High topography is also created by certain etching processes, such as V-groove and trench etching that result in steep-walled features. Patterning submicron-sized features across this topography, and achieving the necessary line-width control, is extremely difficult. The difficulty lies in patterning submicron geometries on top of, and through the recesses of, the rugged IC topography. Lithography with multilevel interconnection processes could be more reliable if a planar (or quasi-planar) surface could be provided. This would result in patterning on a smooth, flat surface, where line sizes are much easier to control.

Several methods can be used to planarize the metallization and properly fill via holes. Bias sputtering, tungsten CVD, and laser melting (or laser planarization) are typical examples. Laser planarization requires the fewest steps, but technologically is the least understood. In this section, we will discuss the origin of high topography structures and the resulting problems for both lithography and interconnect metallization. We will focus on the use of lasers for surface planarization as a potential solution to topography problems.

7.7 IC Topography Problems

The evolution of the integrated circuit has been characterized by a steady reduction of circuit line widths, allowing more and more functions per unit area. Line widths have been reduced from several microns in width to several tenths of a micron, approaching 0.15 μm for advanced IC devices (1). Table 7.1 shows IC device trends over a several year period, indicating the rapid reduction in resolution and increase in the number of layers needed to fabricate an integrated circuit. As the IC becomes more three dimensional, higher aspect ratio structures are generated, and the need for surface planarization increases.

The theoretical resolution limit for optical microlithography is estimated to be in the range of 0.1-0.2 μm for volume-manufactured devices. This assumes the use of optical lithography at a 193-nm laser source with a high numerical aperture (0.60 N.A.) lens and a K factor of 0.50. The thickness of the patterned semiconductor film layer must remain relatively constant in order to preserve high speed signal propagation and distribution of power. Integrated circuit film thicknesses have been reduced to meet some of the needs of more advanced devices, and typical thicknesses range from 0.6 to 1.5 μm, depending on the function of the particular layer. Gate oxide thickness

Lithography Parameters Change By Exposing Wavelength

Wavelength (nm)	Resolution (um)	k_1 Factor	Numerical Aperture
436	0.70	0.80	0.50
365	0.51	0.70	0.50
248	0.32	0.65	0.50
193	0.21	0.60	0.55
157	0.14	0.55	0.60

Table 7.1: Integrated Circuit Device Trends

can be as low as 100 Å. IC film thicknesses below a critical level, especially interconnection metal layers, will not retain the necessary electrical resistance to support device specifications.

The continued reduction in linewidth, with a constant layer thickness, creates the need for multiple innerconnection layers to allow an increased number of functions per unit area. Advanced IC chips may contain as few as 1-2 and as many as 6 or more individual innerconnection metal or silicide layers. Linewidths of 0.25 μm, etched in layers 1.0 μm thick, result in high aspect ratio (4:1) structures. These structures are etched anisotropically on the surface of silicon wafers using special reactive ion etching (RIE) and similar reactive ion plasma methods(2).

The close spacing of these geometries and the relatively deep level of the etching create sharp cornered structures. When overcoatings of oxides or aluminum metallization are applied to these structures, the deposited layer frequently does not cover the corners or coats so thinly on the corners that they subsequently break down during etch operations, causing reject chips. Figure 7.6 shows a cross-sectional view of the various layers of a device and typical topography.

7.8. RESIST IMAGING ON TOPOGRAPHY

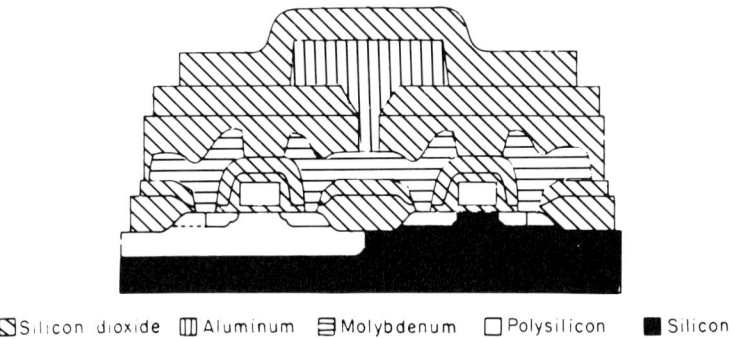

Figure 7.6: Cross-Sectional View of an IC Device

7.8 Resist Imaging on Topography

The variation in thickness of the photoresist covering IC topography (steps) can be as high as 10:1, with only 1500 Å of photoresist on the top of the steps, and 15-20,000 Å of resist in the trench between etched structures. Imaging the same linewidth in a resist with significant thickness variation is difficult and causes problems in maintaining constant linewidth across the IC surface.

The etched structures shown in the SEM photos in Figure 7.7 are spaced closely to preserve space, allowing for an increase in the density of the device. The etching process is largely anisotropic or as close to anisotropic as can be achieved for a given set of etch conditions. A slight wall angle (5-15 degrees) is used to ensure more uniform coverage of the next deposited layer (metallization or dielectric) or to make patterning the resist easier.

In addition to providing uniform, void-free coverage of these sharp-cornered, high aspect ratio structures during oxide and metal deposition, problems can occur in patterning lines of consistent width across these structures and in depositing dielectric and metal layers into submicron-sized via openings for multilevel interconnection.

The variation in resist thickness across these steps results in a tendency

Figure 7.7: Photoresist Imaging over High Topography

to overexpose the thin resist layer on the top of the steps in order to deliver sufficient energy to adequately expose the thick portion. The result is a change in the line width, which may cause a variation in the electrical performance of the device and cause a reject chip.

A possible solution to the problem of imaging over topography is to use resist materials that have high contrast or gamma, defined by the formula for contrast, as follows:

$$Contrast = \left\{\frac{Imax + Imin}{Imax - Imin}\right\} \qquad (7.3)$$

Sufficient contrast in a resist material may not be available, in which case planarization is used so solve the problem of imaging over high topography. Higher gamma value resists are being researched, with the goal to produce contrast values as high as 10. This class of resist materials will be used with surface layer imaging processes. Current gamma values for production resists are in the range of 2.0-5.0.

A high contrast resist would have a gamma value of 8-10, but would still need to retain all of the other resolution and wide process latitude properties necessary for high resolution, production lithography. Formulating a resist material of this description is not simple. Planarization relaxes the demands on the resist imaging properties, especially in thicker layers. Planarization processing with polymers entails placing a much thinner resist imaging material on the surface of the planarizing layer. Imaging into a very thin (0.1-0.5 μm) layer permits much higher resolution capability, but is typically not used unless high topography or high reflectivity problems exist.

7.9 Planarization with Polymer Coatings

Planarization is a flattening or smoothing out of the wafer surface topography by 1) filling in the deep "trench" areas; 2) etching the top surface of an etched structure; 3) filling in via holes; or 4) some combination of these. Planarization modeling is used to map optimal parameters for layer thickness. Modeling permits the process engineer to design a parameter set that eliminates the common problems associated with wide excursions in film thicknesses over etched steps. A model (4) used to characterize IC topography variation and planarization processing is given in Figure 7.8.

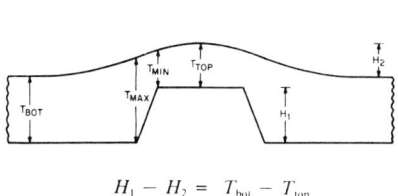

$$H_1 - H_2 = T_{bot} - T_{top}$$

where H_1 = height of the topographical feature
H_2 = amount of polymer film thickness
T_{bot} = thickness of the polymer from the substrate to the top *without* influence of the step (the thickness taken from a resist supplier spin speed versus thickness chart, called spun-on thickness)
T_{top} = polymer thickness on the top of the topographical feature

Figure 7.8: IC Topography Modeling

Note that in this calculation "T_{max}" equals the area of greatest polymer thickness, always taken at the base of a step, and "T_{min}" equals the area of least polymer thickness, always taken at the edge of a step.

Planarization with polymers is accomplished by coating the topography with a resist formulated for this purpose, usually a high solids-content formula with excellent flow properties to wet and flow resist into the deeply etched trenches or small, high aspect ratio areas.

After coating the resist and allowing it to flow into the topography, the resist is baked. The wafer is then etched with oxygen in a dry atmosphere(oxygen plasma or reactive ion etcher) to remove excess resist. Subsequent etching of the resist and underlying oxide or other semiconductor material results in a reduction in topography, or planarization.

The gas mixture used is varied according to the material being etched. Oxygen is used for polymer etching, and other gas mixtures are used to etch oxides, nitride, aluminum, and other films.

7.10 Nonlaser Thermal Planarization

Planarizing can also be accomplished by heating and reflowing low-melting point glasses, such as borophosphosilicates (BPSG) and phosphosilicates (PSG). These are dielectric layers that are doped with varying levels of impurities to lower their melting point. The typical melting point range for these glasses is 750-950°C.

Planarization of insulating layers is also accomplished by rf-bias sputter deposition. The nonlinear rates of deposition over topography serve to self-level the surface. Thermal reflow and sputter deposition have problems of their own. Thermal reflow requires high temperature processing which will increase wafer nonflatness. The high temperature may also change junction depths. Sputtering is both thermal and defect generating, and adds cost and complexity to the overall manufacturing process. Simple processing with low thermal budgets is more desirable.

7.11 UV Laser Planarization

Excimer and other UV and non-UV lasers for direct planarization of metal and dielectric layers in IC fabrication have been demonstrated. Direct laser planarization offers some advantages over the polymer and thermal methods discussed earlier.

7.11. UV LASER PLANARIZATION

A unique property of a UV laser, for example, is that high intensity UV energy can be localized in the top 1200 Å, thereby keeping heat from entering the bulk of the wafer. A laser beam and appropriate beam delivery optics are used to planarize the wafer surface by locally melting and reflowing metals and dielectrics, with minimal process interruption.

Another benefit of UV laser planarization is simple processing. A single step method avoids process interruption, reduces handling, and increases process yields compared to multistep processes. Most other planarizing techniques involve multiple process steps.

As a wafer progresses through the manufacturing process, IC topography increases. By the time the wafer reaches the interconnection mask level, when conductive pathways are etched on the surface to interconnect all the devices created in the bulk of the wafer, considerable topography has been generated. The interconnection materials for ICs are aluminum, aluminum alloys, and silicides; other metals used for similar functions in electronics include copper, silver, and gold. These metals will fill the "via" holes, making ohmic contact with the silicon exposed below. Laser planarization not only melts the surface, causing some leveling, but also flows molten metal down into the tiny via openings which are normally very difficult to fill.

7.11.1 Processing

The rapid transfer of thermal energy from a source into a thin metal or dielectric film, without conduction to the underlying layer, is the reason for using an excimer laser to planarize in IC fabrication. The duration of an excimer laser pulse is in the 15- to 25-nsec range. The entire reaction kinetics involve several stages (irradiation time, heating, melting, flow, thermal equilibrium, cooling) but take place in less than a few microseconds.

In a conventional thermal reflow furnace, wafers are exposed to melt times in excess of 100 μsec, resulting in melting of underlying layers. For example, silicon dioxide will react with molten aluminum, forming silicon and volatile by-products. Dielectric layers, used to prevent metal diffusion into active silicon regions, can melt into the molten metal layers if heating times exceed even 1 μsec.

Thus, submicrosecond heating is desired in most IC processes to prevent these colateral reactions and to preserve the integrity of the IC layers and diffusion areas beneath the layer being planarized. Integrity means preservation of morphology, crystal orientation, diffusion depth, grain structure, and stoichiometry.

7.11.2 Laser Planarization Reactions

High purity copper, gold, and aluminum, in the liquid state, are characterized by relatively low viscosities and high surface tension. These properties allow the metals to planarize rapidly, in fractions of a second. Further, the high rate of thermal diffusion in these materials (1 cm^2/sec for aluminum) permits rapid energy transfer of the laser pulse through the layer of metal. Typical metal layer thicknesses are up to 1-2 μm, resulting in complete heating of a layer in less than 10 nsec in the case of aluminum.

The surface tension of these interconnect metals (aluminum, copper, gold) is extremely high in the molten state, typically 50 times that of water (5); also, their molten-state viscosity is low relative to water. These physical properties cause the liquid metals to planarize very rapidly. Mathematical models of transient fluid flow only partially explain the submicrosecond behavior of molten metals in laser planarization reactions.

The low thermal diffusivity of amorphous silicon dioxide (0.009 cm^2/sec) makes it an ideal thermal barrier (for 1-μsec pulses) in layer thicknesses down to 9000 Å (5). Key physical properties of the metals and dielectrics used in IC interconnection technology are listed in Table 7.2.

SUBSTRATE TEMP.	LASER SPOT			
	MELTING	PLANARIZ.	VIA-FILL	ABLATION
25°C	>(2.4 mm)^2	(1.9 mm)^2	(1.5 mm)^2	(1.3 mm)^2
200°C	≥(2.8 mm)^2	(2.1 mm)^2	(1.8 mm)^2	(1.6 mm)^2
350°C	>(2.8 mm)^2	(2.3 mm)^2	(1.9 mm)^2	(1.7 mm)^2
450°C	>(2.8 mm)^2	(2.6 mm)^2	(2.1 mm)^2	(1.9 mm)^2
SUBSTRATE TEMP.	LASER FLUENCE			
	MELTING	PLANARIZ.	VIA-FILL	ABLATION
25°C	<1.75 J/cm^2	2.40 J/cm^2	3.21 J/cm^2	3.76 J/cm^2
200°C	≤1.40 J/cm^2	2.10 J/cm^2	2.57 J/cm^2	2.99 J/cm^2
350°C	<1.40 J/cm^2	1.85 J/cm^2	2.40 J/cm^2	2.76 J/cm^2
450°C	<1.40 J/cm^2	1.56 J/cm^2	2.10 J/cm^2	2.40 J/cm^2

Table 7.2: Planarization Parameters for Laser Via Filling at 308 nm (Courtesy of R. Liu, AT&T Bell Labs)

7.11.3 Laser Planarization: Reaction Components

Planarizing a layer of metal or dielectric on a semiconductor or other substrate involves a complex set of interrelated reactions. The major components of this complex reaction are 1) the laser beam; 2) the substrate tem-

7.11. UV LASER PLANARIZATION

perature; 3) thickness of layer being planarized; 4) optical properties of the layer being planarized; and 5) morphology of the underlying layer. These topics are addressed in this chapter.

7.11.4 The Laser Beam

The laser beam is a major factor in determining planarization results and can be modeled in optimizing the planarization reaction to produce repeatable results. The beam is optically homogenized to remove "hot" spots, or areas of nonuniformity. Beam conditioning or homogenizing is necessary to protect the beam delivery optics and optical coatings from high fluence areas or hot spots in the beam. Regions of nonuniform, high photon concentration will degrade the coatings or the lenses; this can lead to solarization and physical damage (due to heating and expansion) inside the optical element.

Beam diagnostic equipment is used to map the distribution of energy across the beam axes and to measure the fluence at various points in the optical path. The extent of beam homogenizing can be quantified with the proper beam metrology equipment. Pin diode mapping and large area detector arrays are typical devices used. It is critical to determine the laser fluence throughout the optical path, down to the target of substrate.

7.11.5 Laser Fluence

Fluence determination is an important factor in identifying the necessary energy to initiate planarization. Planarization reactions have a calculable melt threshold, and the laser system is operated slightly above this threshold. In the case of aluminum, a fluence in the range of 3.5-6.0 J/cm^2 is required for complete planarization of a 1-μm-thick layer (6). This energy level is needed because of the inherent high reflectivity of aluminum, generally in the 90-93% range, depending on the amount of copper and silicon present in the aluminum.

The melting of the aluminum will occur very rapidly under the energy levels cited above and may not achieve sufficient flow to properly planarize the underlying layer, especially in the case where deep via holes need to be completely filled with molten metal. Additional flow is achieved by heating the wafer (or other substrate) to temperatures in the 200-400° range. The heated substrate will prolong the time the aluminum is in the molten state, allowing the metal to fill small holes and flow into narrow valleys on the wafer surface. Figure 7.9 shows planarization of aluminum.

CHAPTER 7. ANNEALING AND PLANARIZING

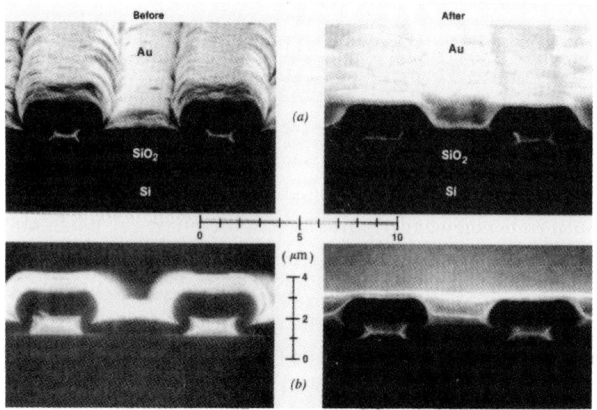

Figure 7.9: Planarization of Aluminum

The minimum beam fluence required to planarize aluminum and other metals is partially a function of reflectivity properties. Reducing the reflectivity of metals with antireflection coatings increases the coupling of energy into the metal layer and thereby reduces the required minimum dose required to achieve planarization.

7.11.6 Antireflection Coatings

Antireflection coatings based on organic polymers typically break down during medium (2-3 J/cm^2) to high (4-10 J/cm^2) fluence exposure. Layers of copper (10), silicon (5), and titanium (7) provide greater resistance for high fluence applications and have been used to provide significant reduction in UV reflectivity and can increase the efficiency or throughput of a planarization process.

Organic coatings can be used for lower fluence applications and are spin coated or vacuum deposited primarily on the surface of the metal layer. Aluminum contact layers on wafer substrates typically traverse a wide variety of other reflecting dielectric and metallic portions of the wafer, including polysilicon and aluminum. Reflection can be severe across the varying wafer topography, with some areas providing angled sidewalls that act as mirrors

7.11. UV LASER PLANARIZATION

for incident radiation.

7.11.7 Laser Planarization Equipment

An excimer laser system for planarization of metals consists of several subsystems. These include the excimer laser, beam delivery optics, and a stage to provide movement of the sample under a fixed beam, movement of the beam across a fixed sample holder, or both, a computer control system, beam homogenizer, video camera system for viewing, beam focusing and shaping optics, turning mirrors, and beam attenuator optics. A schematic of a typical laser planarization system is shown in Figure 7.10.

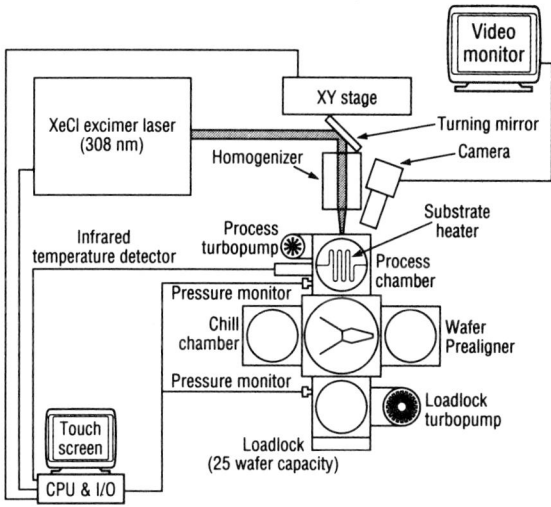

Figure 7.10: Schematic Diagram of a Laser Planarization System

7.11.8 Laser Source

The laser for the planarization system shown is a 300-Hz, xenon chloride (XeCl) excimer laser. The pulse energy for this system is 500 mJ, with a pulse duration of 45 nsec. The beam is homogenized from a quasi-gaussian shape to a "top-hat" shaped beam, as shown in Figure 7.11. The uniformity of the beam after passing through the homogenizer is +/- 5%; the "a" example is the raw beam, and the "b" example is after being homogenized.

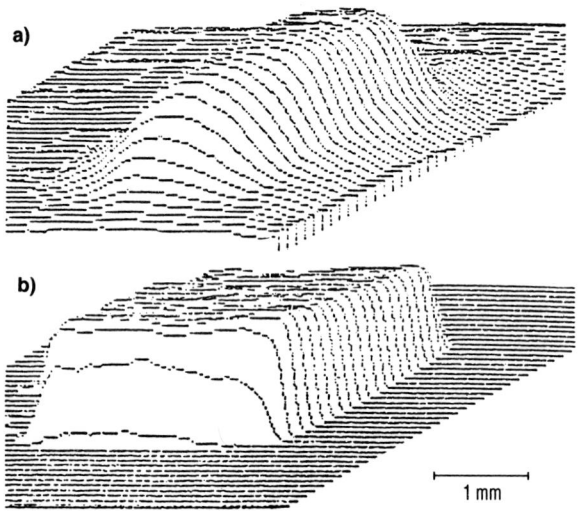

Figure 7.11: Laser Output Beam Profile

The variety of film types (metals, semiconductors, dielectrics) used in IC fabrication have a wide range of ablation thresholds and melting points. This creates the requirement that any planarization system be flexible enough to provide a wide range of process parameter controls. For example, the laser reprate should be variable, under computer control, from 1 Hz to the laser maximum, typically 500-1000 Hz.

Other controls include voltage, attenuation, pulse train programming, and variable beam-sizing apertures. Laser voltage is one variable used to adjust the output power of the laser. Beam attenuators are used to selectively remove precalculated amounts of laser energy. The computer system software can allow for exposing a preset number of laser pulses or "pulse trains," based on the energy requirements of the application. Up to 1000 pulses (or more) can be programmed into a single burst.

7.11.9 Beam Shaping and Masking

Controlling the size of the irradiated area on the sample can be done with fixed or movable apertures, or with prepatterned chrome-on-quartz masks. Fixed apertures are be manually or automatically inserted in the beam path; movable, bladed apertures may also be driven manually (micrometer) or au-

7.11. UV LASER PLANARIZATION

tomatically (stepper motor). Patterned chrome on quartz masks is inserted in the beam bath for multiple exposures of the same pattern or for superimposing various patterns on top of each other.

Shaping the beam permits on-the-fly laser patterning. The operator can program x,y,z, and theta motions in the computer, executed by the stages in the laser patterning system. In deep layer micromachining, computerized control over beam fluence (autoattenuation) and z-motion stages allow the operator to cut perfectly vertical shapes deep in plastic (vias on ICs) or other low density materials such as photoresist, polyimide, and oxides. Without z-tracking in high aspect ratio structures, the laser beam will become shadowed as the hole is deepened, reducing the beam fluence and changing the cutting rate and shape of the sidewall. High pressure (or vacuum) nozzles are used to remove the ablation by-products.

7.11.10 Equipment Subsystems

Equipment for laser planarization may also include any of the following subsystems: a) robotic substrate handling; b) various types of nozzling systems to remove debris or gaseous by-products; c) vacuum cells through which the beam is sent for laser-assisted deposition and etching; d) special wafer chucks for heating or cooling the parts during laser irradiation; and e) autofocus systems for tracking the beam in the "Z" plane, as into via holes or for deep micromachining.

7.11.11 Beam Delivery Optics

UV laser planarization reactions involve relatively high energy levels being delivered to the substrate in relative short (1 μsec) timeframes. Delivery systems integrate dose and pulse metrology tools to monitor the pulse profile and energy. The system computer control will integrate this data and calculate the total dose for a given reprate. Figure 7.12 shows a schematic of an optical system used for UV laser planarization of gold and aluminum alloys (6).

As the beam exits from the laser, it passes through an interlocked shutter. Interlocks are standard industrial safety features on all laser delivery systems. The shutter subsystem can be programmed to deflect the first one to two pulses; this is done with excimer lasers, as these first pulses may be low in energy or atypical due to initial electrical discharge effects.

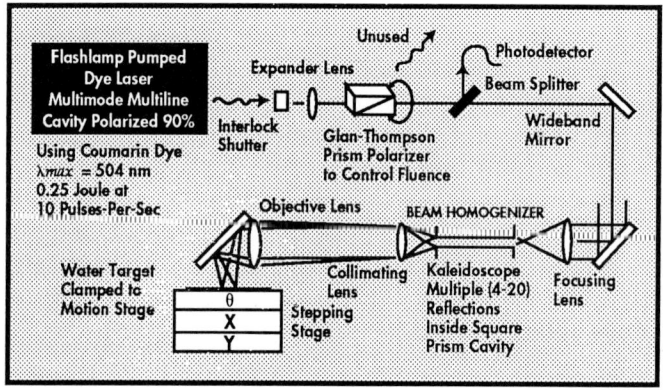

Figure 7.12: Planarization System Optics

Expanding Lenses and Beam Splitters

A key optical element in the delivery system is an expanding lens to reshape the beam from a rectangular to a square shape since raw beams from excimer lasers are rectangular. A prism polarizer is used for attenuation. A beam splitter is then used to pick off a small portion (2-10%) of the beam to sample the profile before passing the energy across two additional beam-turning mirrors and into a focusing lens.

Homogenizers

The homogenizer is an important component in any laser planarization system, principally because high quality laser processing requires very uniform energy. The extreme thinness (200-1000 Å) of films used in IC fabrication makes these structures sensitive to hot spots in a laser beam. The nonuniform hot spots penetrate the layer being melted and ablate areas beneath the layer being planarized. The specification for thickness variation for some films is very tight and is required to preserve the electrical parameters of the IC. Laser beam energy must also be highly controlled and uniform in these applications.

A typical design problem in building a laser homogenizer is the trade-off

7.11. UV LASER PLANARIZATION

between beam uniformity and beam energy. A mirror-tunnel homogenizer that bounces the beam over four mirrors (arranged in a square shape, facing into each other) provides high uniformity, but sacrifices over 50% of the beam energy. Simpler beam-folding homogenizers are much less lossy, but only provide +/- 15% energy uniformity. Most semiconductor applications require +/- 2-5% beam uniformity or better.

A kaleidoscope has between 4 and 20 reflections inside a square prism cavity. This type of homogenizer will produce square spots, useful for irradiation of an individual die on a silicon wafer.

The kaleidoscope homogenizer is capable of mixing laser light to reach a uniformity of about +/- 5% at the wafer plane. A final collimating lens and an objective lens complete the optical train before delivering the beam past the turning mirror and onto the substrate.

Beam Sizing

The laser beam can be optically sized according to the application. Beam size can be controlled by apertures, pre-etched masks, or by the homogenizer, variable from 2 to 6 mm^2 in both X and Y directions. Changing the beam area will also change the beam fluence for a given input energy.

If the size of the beam is insufficient to cover a single die site on the wafer, a means of generating multiple exposures (field butted or overlapping) is needed. To accomplish this, a grid of overlapping exposures is generated with a scanning stage. The beam is scanned across the surface of the wafer at a rate and fluence sufficient to planarize the metal layer.

7.11.12 Applications

Metal Interconnection Layers

There are several metals used or interconnection layers in electronic circuits that require or can benefit from planarization in multilevel IC interconnection schemes; aluminum and aluminum alloys are commonly used. Other interconnect metallization schemes include TiW alloys, metal silicides, and, in the case of multichip modules and printed circuit boards, copper. These metals are placed over various dielectrics, and the need to planarize a metal layer is a direct function of the topography of the dielectric. In many cases the need to planarize is reduced or eliminated by creating a sloping sidewall instead of leaving a vertical "cliff" for the metal to cover. Tapering the wall angles in IC devices by selecting special etch parameters is used to reduce the severity of the wafer topography. Extensive wall angle tapering (720°) is undesirable since additional surface area is consumed in order to accommodate the expanded base of the tapered structure. Also, it may be easier to use a laser planarizing step than to introduce one or more steps to reduce the severity of the underlying topography.

A high level of planarity can be achieved by planarizing the interconnection layers. Laser planarization of nonrefractory metals provides a way of leveling metal over insulating layers without affecting the underlying layer due to the very localized and rapid surface heating and melting provided by the laser pulses (5).

Long wavelength lasers as well as furnaces (infrared, convection) are used to melt metallization layers, but transfer significant amounts of heat to the underlying layers. Thermal distortion of the wafer from this transferred heat causes pattern distortion in lithography.

Wafer heating below the surface (in the bulk) may also change IC junction depths. The amount of heat necessary to cause this critical problem is different for different parts of the process. Temperatures up to about 200°C are generally acceptable throughout the process. Early in the IC manufacturing cycle, before many junctions are formed, temperatures up to 400°C may be acceptable. In general, high temperatures are undesirable since they cause wafer nonflatness.

Gold Interconnection

In gold interconnection schemes, an adhesion layer, typically 250-400 Å thick, is used at the gold/oxide interface. Adhesion layers are thin films of tungsten, niobium, chrome, or titanium. The gold layers are generally about 1 μm-thick. This type of metallization scheme can be planarized with a microsecond pulse at a fluence of 1 J/cm^2 at a wavelength of 504 nm. Less fluence will be required with an excimer laser due to the higher energy per photon at 308-, 248-, and 193-nm excimer laser wavelengths.

The presence of a dielectric layer under the metal layer increases the latitude available for fluence variation. The dielectric layer acts as a buffer zone, absorbing energy below its ablation threshold. As much as a 2:1 variation in fluence has been used in gold planarization experiments, with good melting, flow, and step coverage of the metal (5). Wide process latitude helps to ensure a high throughput, high yielding manufacturing process.

7.11. UV LASER PLANARIZATION

Planarization of Via Holes

Another major IC application for laser planarization is the filling of via holes. Via holes, through which aluminum or other metallization makes ohmic contact to the silicon substrate, are difficult to fill due to their submicron size and high aspect ratios. For example, a via may be 0.5 μm wide and 5 μm deep (1:10 aspect ratio).

In the metallization step, the conductor (aluminum) is deposited by magnetron sputtering or evaporation directly over the pre-etched vias. Getting the metal to cover the steep sidewalls of the submicron via holes is difficult and can lead to poor reliability of the device. The sidewall of the via can "shadow" the metallization as it is being deposited.

Laser planarization is used in this application to melt the overlying, as-deposited metal, allowing it to flow into the via hole. In research experiments with laser filling of via holes, a thin coating of silicon is typically necessary as an overcoat to increase optical absorbance and be an antireflection coating. However, during planarization and melting, the silicon alloys with the aluminum, causing redeposition of silicon into contacts and increased contact resistance. A direct, one-step planarization is therefore most desirable and will be discussed later in this chapter.

Via holes in advanced IC devices are typically small with vertical or near-vertical sidewalls in order to conserve space. The via connects (makes physically and electrically ohmic contact) the aluminum or aluminum alloy to the silicon substrate. The sharp corners and vertical sidewalls make step coverage of overcoated metallization a problem. Figure 7.13 illustrates the process of via hole planarization.

The via hole structure in this case is a silicon substrate overcoated with an insulator consisting of 1-μm-thick phosphosilicate glass (PSG). A 1-μm thick layer of pure aluminum is magnetron sputtered over the insulator (8).

Filling of the via by laser planarization can be performed using a single 193-nm excimer laser pulse. A simple focusing lens is used with the laser to increase beam fluence, which is about 10 J/cm^2. The laser fluence is calculated by dividing the focused beam size into the pulse energy. The melt area used was 0.8 mm vertical by 2 mm horizontal.

Note in Figure 7.13 that the aluminum is only partially planarized with 3 J/cm^2; the full fluence of 10 J/cm^2 is needed to fully planarize the layer. The melting at low fluence draws the metal to the via, but leaves it relatively uneven. The full melting of the metal causes mass transport into the high aspect ratio via hole. In this case, the hole was 0.6 μm in diameter and 0.7

CHAPTER 7. ANNEALING AND PLANARIZING

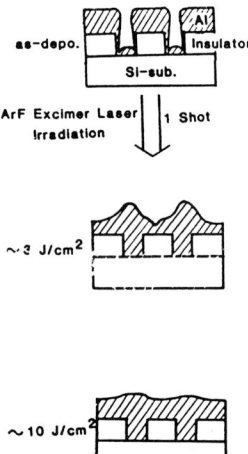

Figure 7.13: Via Hole Laser Planarization Process

μm in depth (1.2 aspect ratio). When the fluence is increased to 15 J/cm^2, the aluminum is ablated.

The aluminum in the above experiment was a single-crystalline grain connecting to (110) silicon. Solubility of silicon in aluminum typically causes junction spiking during furnace alloying cycles (425°C, 10 min), a problem for thin junctions. Laser planarization avoids this problem due to the ultra-short melt times during planarization, preventing silicon/aluminum alloying from occurring.

Planarization of the dielectric layers of silicon dioxide requires 1-2 J/cm^2 fluence, but due to the depth of the via (2-3 μm) and the increased thickness of metal in the filled via, a fluence of 3-5 J may be needed to fully planarize the structure. This fluence level is high but is still below the ablation threshold for these films. Due to inherent nonuniformities (hot spots) in the beam, it is advisable to irradiate with a fluence that is at least 2 J/cm^2 below the damage threshold. Figure 7.14 shows a planarized via structure.

Aluminum Interconnect Planarization

The basic problem of poor metal coverage of metal contacts interconnection schemes on advanced IC devices is illustrated in Figure 7.15.

7.11. UV LASER PLANARIZATION

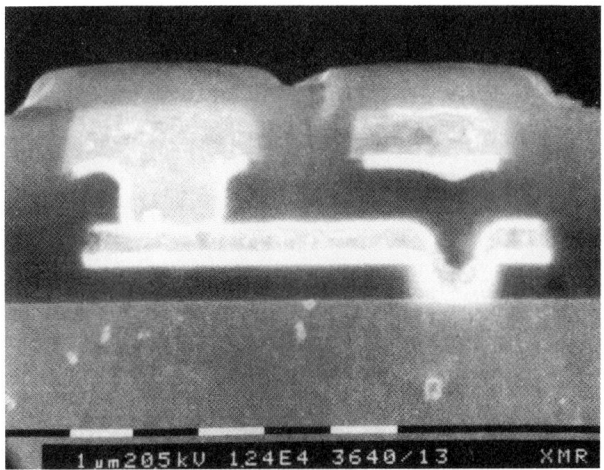

Figure 7.14: Planarized Via Structure

The distance across the via at the bottom, where metal-metal contact is made, is less than 1 μm in most cases; in advanced IC designs, this dimension can be less than 0.5 μm. Note the variation in aluminum layer topography before and after irradiation with the excimer laser. A single laser pulse was used to planarize this layer, using a fluence of 5-10 J/cm^2 (8).

The main steps used to planarize aluminum at 193 nm are outlined in the following section. An argon fluoride excimer laser source was used. The pulse duration was 15, with a 170-mJ pulse energy and a reprate of 1 Hz. The speed of heating with 193-nm radiation greatly exceeds the 1-μsec melt time when using a dye laser in the same application and eliminates the undesirable step of overcoating the aluminum with silicon for increased absorption. At 193 nm, aluminum absorption is approximately 1.3 x 10^{6}cm^{-1}; reflectivity is approximately 90%.

193-nm Aluminum Planarization Process

The following process parameters have been used to planarize aluminum metallization on IC structures at 193 nm:

1. Aluminum Deposition: A 0.5-μm-thick film of aluminum, alloyed with 1% silicon, is deposited by magnetron sputtering over 1-μm thick PSG

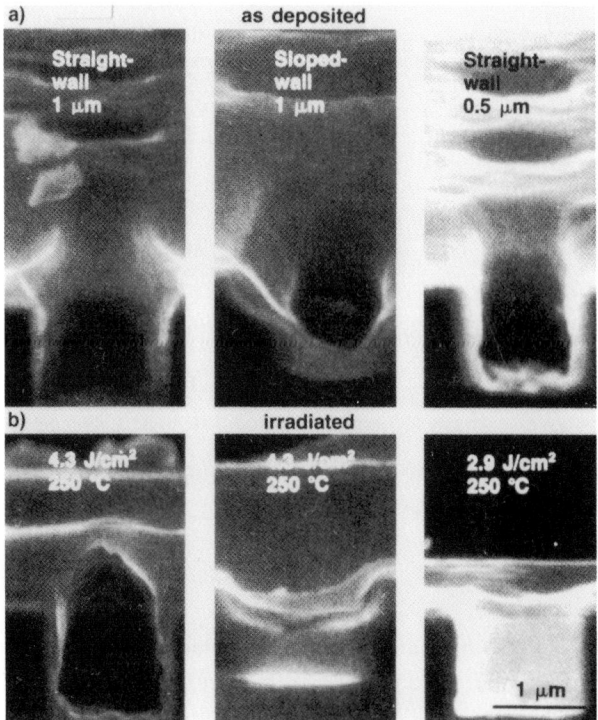

Figure 7.15: Planarizing of Aluminum

over silicon.

2. Patterning: The photoresist is coated, baked, exposed, and developed on the surface of aluminum.

3. Oxide Deposition: A double layer of 100 nm of silicon dioxide and 600 nm of PSG is deposited as an insulating layer. This step can be eliminated if short melt times are used.

4. Patterning and Etching: Via hole patterns are formed and subsequently etched. These permit electrical connection between the metal interconnection layers.

5. Metalization: A layer of pure aluminum (second layer) is sputter-deposited.

7.11. UV LASER PLANARIZATION

6. Laser Planarization: The upper metal layer is irradiated with the 193-nm laser pulse, causing melting and leveling of the metal.

The extremely short melt times possible with the 193-nm excimer laser avoid the problem of melting the underlying layers. The only significant change in the metal layer after planarization is in grain size; as-deposited aluminum has a grain size of 0.5 μm, where 193-nm reflowed aluminum has a grain size of about 10 μm.

When the metal is not completely planarized, and thickness variations occur in circuit lines, dry etching with very high (20:1-30:1) selectivity ratios is used. This permits etching through the thickest areas of the metal (or polysilicon) without changing the lateral dimension of the thin areas of the layer that are being significantly overetched. It is obviously important to keep the etch direction vertical (anisotropic etching) to prevent any lateral etching

Complex CMOS devices may have as many as 7-8 interconnection metal layers. Without planarization, variations in the height of the metal will increase over several layers to the point where it becomes resolution-limiting. Laser planarization not only increases the latitude in the patterning process, but also permits higher resolution. One of the challenges that remain for interconnection planarization is providing the same latitude in the laser process so that changes in substrate reflectivity, metal thickness variations, and laser energy variations will not cause significant changes in the planarization behavior.

Interconnection Planarization Parameters

The required beam fluence for planarization varies widely according to the types, combination, and thicknesses of films being irradiated. Interconnection technology for electronics devices is fabricated in several ways. These include: a) aluminum/aluminum alloy metal for basic IC interconnection to active silicon layers; b) gold conductors to form interconnection on thin film devices, IC packages, and custom solid state devices integrated onto this film circuit; and c) copper films for printed circuits and as conductive layers between polyimide layers in multichip modules and other high density IC interconnection schemes used in wafer scale integration.

Other metals and nonmetals that can be planarized include chromium, titanium, metal and silicon oxides, polyimide, silicides, nitrides, and other thin seminoble and noble metal layers. Each type of material is used at varying thicknesses and in various combinations.

Aluminum is especially difficult to planarize due to its high reflectivity and refractory native oxide. The native oxide remains at temperatures well above the aluminum melt point and must be sputtered off, replacing it with a thin (150 Å) protective layer of amorphous silicon dioxide. This thin coating of silicon dioxide is also used as a protective layer on mirror optics used to deflect excimer laser beams in high fluence micromachining applications. The oxide also serves as a protective layer which can be cleaned or dusted periodically. Without the protective oxide coating, the bare metal aluminum mirror surface would be scratched during cleaning; if the aluminum was overcoated with an optical (AR) coating, it also would be subject to mechanial damage.

In planarization reactions, the silicon oxide overcoat becomes alloyed with the aluminum during the melting phase of planarization. The alloying (e-fold decay depth of 0.3 μm) results in a slight loss of reflectivity of the aluminum (5). The planarization of the aluminum can take place at ambient temperature, without vacuum. The small amount of aluminum oxidation that occurs during melting is insufficient to impact the planarization process.

ASIC Devices

Another application for aluminum planarization is in ASIC (application specific integrated circuit) chip fabrication, where the final aluminum layer is deposited and left undefined. The final personalization of this layer is left until the specific application for which the chip is intended is decided, and the aluminum is then planarized, patterned, and etched. The wafers are loaded into a sputtering chamber to remove the native oxide, then laser planarized, and sent into the photolithography process for resist coating and patterning. Figure 7.16 shows an optical system for planarizing, exposure, or deposition of material.

If a polysilicon interconnect layer resides below the aluminum, fluence levels below the damage threshold of the polylayer must be used. This is generally not a problem when the laser beam is properly homogenized.

7.11.13 Planarization-Induced Changes in Thin Films

Laser planarization will bring about some changes in the properties of thin films, such as electrical resistance, grain size, and reflectivity. Table 7.3 shows the electrical parameters for a 64K SRAM device before and after laser planarizing. Note that the distribution of contact resistance values is

7.11. UV LASER PLANARIZATION

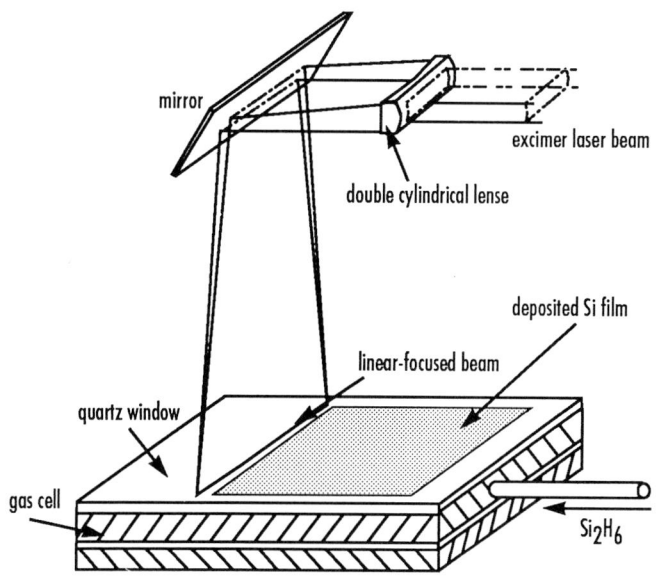

Figure 7.16: Scanning UV Exposure/Deposition System

slightly better with no significant shift in threshold or breakdown voltages.

Electrical Results of Laser Planarized 64K SRAM

Parameter	Control	Planarized
Gate width(μm)	0.79 ± 0.06	0.82 ± 0.06
V_T n-trans (V)	0.69 ± 0.01	0.68 ± 0.06
V_T p-trans (V)	− 0.91 ± 0.01	− 0.92 ± 0.01
R_C n$^+$ (ohm/contact)	12.43 ± 2.55	12.4 ± 0.2
R_C p$^+$ (ohm/contact)	38.08 ± 1.30	39.5 ± 0.64
R_C poly (ohm/contact)	1.36 ± 1.27	0.99 ± 0.05
BV n$^+$ (V)	14.10 ± 0.00	14.10 ± 0.00
BV p$^+$ (V)	20.70 ± 0.14	20.82 ± 0.12

Table 7.3: Electrical Results of Laser Planarized SRAM

Resistivity measurements shown in the table indicate a very small change in resistivity caused by planarization (5). In some cases, the resistivity value increases, and in others it decreases. Additional data are needed to conclude a predictable change in electrical behavior arising from laser planarization processing.

7.11.14 Conclusion

Data presented in this section show the applications for laser planarization in the fabrication of high density integrated circuits. Specifically, laser planarization greatly reduces the problems caused by high aspect ratio topography and sharp cornered structures, while at the same time making it possible to add additional layers to a device and continue with the same processing (deposition, patterning, and etching) technology.

Future devices will require additional layers in order to achieve the required resolution and density, and planarization can play a role in facilitating the transition to increasingly complex chips while maintaining much of the same process technology.

References

1. Sparkes, C., DiSessa, P., "Technical Advances and Applications for Laser Microlithography in the 90's," GCA, A Unit of General Signal, Andover, MA.

2. Elliott, D.J., "IC Fabrication Technology " Second Edition, McGraw Hill Book Company, New York, 1989.

3. SEM photo, courtesy of Shipley Company, Newton, MA.

4. Model supplied by L.K. White, RCA, Princeton, NJ

5. Tuckerman, D.B., Weisberg, A.H., "Laser Planarization," Solid State Technology, April, 1986.

6. Chen, E. Ong, "An Excimer Laser-Based Aluminum Planarization Process," SPIE Vol. 1190, 1989, p. 207-212.

7. Liu, R., Cheung, K.P., Lai, W.Y.C.,"A Study of Pulsed Laser Planarization of Aluminum for VLSI Metallization," VMIC Conference (IEEE), June 12-13, 1989, p. 329-335.

7.11. UV LASER PLANARIZATION

8. Mukai, R., Sasaki, N., Nakano, M., "High-Aspect-Ratio Via-Hole Filling with Aluminum Melting by Excimer Laser Irradiation for Multilevel Interconnection," IEEE Electron Device Letters, Vol. EDL-8, 2 February, 1987.

9. Pramanik, D., Chen, S., "Characterization of Laser Planarized Aluminum for Submicron Double Level Metal CMOS Circuits, 1989 IEDM Proceedings, P. 673-676

Glossary of Terms

Absorption Band The area of the spectrum in which the absorption coefficient is at a maximum.

Absorption Coefficient The internal absorbance of a material; given that the unit transmission of a material is 't', the absorption coefficient (a) is $a = \log_e t$.

Amorphous A disordered, glassy solid state of a semiconductor (or other substance) as compared to a highly ordered crystalline solid state. The thermal, optical, electrical, physical, and other properties vary considerably when comparing the amorphous to the crystalline state of the same material.

Annealing A process wherein a solid material is heated and cooled in order to change its physical, optical, electrical, or thermal properties; UV laser annealing is performed to reverse the damage caused to the crystal lattices of semiconductor layers caused by high energy ion bombardment (ion implantation).

Lattice Constant A length that denotes the size of the unit cell in a crystal lattice. In a cubic crystal, the geometry of each type of material varies, so it must be individually calculated. Laser energy in annealing a crystal makes significant changes to the lattice structure (damage, plane shifts, melting, and resolidification, stress buildup, etc.).

Planarizing To make a surface flat; in a UV laser planarization, heat causes melting and reflow of uneven deposits which tend to self-level when molten.

Via A small opening in an insulative semiconductor layer which is filled with a metallic conductor to permit ohmic contact with the underlying silicon semiconductor device; multiple layers of semiconductor devices are connected with via metallization.

Reading Bibliography

1. Mathe, E.L., Naudon, A., Repplinger, F., Fogarassy, E., "Morphology of $Se_1 - xGe_x$ Thin Crystalline Films Obtained by Pulsed-Excimer-Laser Annealing of Heavily Ge-Implanted Se," Applied Surface Science, Vol. 86, p. 338-45, Feb. 1995.

2. Wagner, F.X., Dhese, K., Key, P.H., Sands, D., Jackson, S.R., Kirbitson, R., Nicholls, J.E., "Photo-Luminescence of Pulsed Excimer Laser Annealed Sb-Implanted CdTe," Applied Surface Science, Vol. 86, p. 364-7, Feb. 1995.

3. Carluccio, R., Stoemenos, J., Fortunato, G., Meakin, D.B., Bianconi, M., "Microstructure of Polycrystalline Silicon Films Obtained by Combined Furnace andLaser Annealing," Applied Physics Letters, Vol. 66, No. 11, p. 1394-6, March 1995.

4. Takarev, V.N., Marine, W., Prat, C., Sentis, M., " 'Clean' Processing of Polymers and Smoothing of Ceramics by Pulsed Laser Melting," Journal of Applied Physics, Vol. 77, No. 9, p. 4714-23, May 1995.

5. Ulrych, I., El-Kader, K.M.A., Chab, V., Kocka, J., Prikryl, P., Vydra, V., Cerny, R., "Properties of Recrystallized Amorphous Silicon Prepared by XeCl Excimer Laser Irradiation," Materials Science Forum, Vol. 173-174, p. 29-34, 1995.

6. Compaan, A., Savage, M.E., Jayamaha, U., Azfar, T., Aydinli, A., "Raman Studies of Doped Poly-Si Thin Films Prepared by Pulsed Excimer Laser Annealing," Materials Science Forum, Vol. 173-174, p. 197-202, 1995.

7. Zhuang, Q., Ishigoh, K., Tanaka, K., Kawano., K., Nakata, R., "Laser Silicon Ablation Studied by Tiem-of-Flight Mass Spectrometry: Generation of Si Ions and Neutrals and Their Translational Properties," Japanese Journal of Applied Physics, Part 2 (Letters), Vol. 34, No. 2B, p. L248-51.

7.11. UV LASER PLANARIZATION

8. Baseman, R.J., Kuan, T.S., Aboelfotoh, M.O., Andreshak, J.C., "Pulsed Laser Planarization of Metals for IC Interconnect," Photons and Low Energy Particles in Surface Processing Symposium, p. 361-9, 1992.

9. Noguchi, S., Kuwahara, T., Kuriyama, H., Hanafusa, H., Kiyama, S.,Nakano, S., "Poly-Si by Excimer Laser Annealing with Solidification Process Control," Transactions of the Institute of Electronics, Information and Communications Engineers C-II, Vol. J76C-II, No. 5, p. 241-8, May 1993.

10. Carey, P.G., Woratschek, B.J., Bachmann, F.E., "Progress Toward Excimer Laser Metal Planarization and Via Hole Filling Using In-Situ Monitoring," Microelectronic Engineering, Vol. 20, No. 1-2, p. 89-106, March 1993.

11. White, L.K., "Planarization properties of resist and polyimide coatings," J. Electrochem. Soc., Vol. 130, No. 7, pp. 1543-1548, 1983.

12. Delfino, M., Reifsteck, T.A., "Laser activated flow of phosphosilicate glass in integrated circuit devices," IEEE Electron Device Lett., Vol. EDL-3, pp. 116-118, 1982.

13. Delfino, M., "Phosphosilicate glass flow over aluminum in integrated circuit devices," IEEE Electron Device Lett., Vol. EDL-4, pp. 54-56, 1983.

14. Delfino, M., "Laser activated flow for integrated circuit fabrication," SPIE-Laser Processing of Semiconductor Devices, Vol. 385, pp. 32-37, 1983.

15. E.R. Sirkin, I.A. Blech, "A method of forming contacts between two conducting layers separated by a dielectric," J. Electrochem. Soc., Vol. 131, pp. 123-125, 1984.

16. Tuckerman, D.B., Schmitt, R.L., "Pulsed laser planarization of metal films for multilevel interconnects," in Proc. 1985 VLSI Multilevel Interconnection Conf. (V-MIC), IEEE Cat. 85CH2197-2, June 1985, pp. 24-31.

17. Tuckerman, D.B., Schmitt, R.L., Proc. 2nd International IEEE VLSI Multilevel Interconnection Conference, p. 24, Santa Clara, CA (1985).

18. Osborne, J., Magee, T., Leung, C.S., Gildea, P., Recent News Paper, 4th International IEEE VLSI Multilevel Interconnection Conference, Santa Clara, CA (1987).

19. Tuckerman, D.B., Weisberg, A.H., IEEE Electron Device Lett. EDL-7, 1 (1986).

20. Mukai, R., Sasaki, N., Nakano, M., IEEE Electron Device Lett., EDL-8, 76 (1987).

21. Mukai, R., Kobayashi, K., Nakano, M., Proc. 5th International IEEE VLSI Multilevel Interconnection Conference, p. 101, Santa Clara, CA (1988).

22. Baseman, R.J., Andreshak, J.C., Ting, C.-Y., Gupta, A.,Materials Research Society Fall Meeting, Boston, MA (1988).

23. Liu, R., Williams, D.S., Lynch, W.T., IEDM, p. 58 Los Angeles, CA (1986).

24. Lai, W.Y.C., Liu, R., Cheung, K.P., Heim, R., "The Use of Ti as an Antireflective Coating for the Laser Planarization of Al for VLSI Metallization," 6th International IEEE VLSI Multilevel Interconnection Conference, Santa Clara, CA (1989).

25. Sparks, M., Loh, E., Jr., J. Opt. Soc. Am., 69, 847 (1979).

26. Fradin, D.W., Yablonovitch, E., "Comparison of laser induced bulk damage in alkali halides at 10.6, 1.06, and 0.69 microns," in Laser Induced Damage in Optical Materials, A.J. Glass and A.H. Guenther, Eds. US GPO Washington, CD: NBS Special Publication 372, 1972, pp. 27-39.

27. Bass, M., Barrett, H.H., "Laser induced damage probability at 1.06 and 0.69 pm," in Laser Induced Damage in Optical Materials, Glass, A.J., Guenther, A.H., eds. US GPO Washington, DC: NBS Special Publication 372, 1972, pp. 58-69.

28. Porteus, J.O., Soileau, M.J., Bennett, H.E., Bass, M., "Laser damage measurements at CO2 and DF wavelengths," in Laser induced Damage in Optical Materials, Glass, A.J., Guenther, A.H., eds. US GPO Washington, DC: NBS Special Publication 435, 1975, pp. 207-215.

7.11. UV LASER PLANARIZATION

29. Newnam, B.E., Gill, D.H., "Spectral dependence of damage resistance of refractory oxide optical coatings," in Laser Induced Damage in Optical Materials, Glass, A.J., Guenther, A.H., eds. US GPO Washington, DC: NBS Special Publication 362, 1976, pp. 292-300.

30. Newnam, B.E., Gill, D.H., "Ultraviolet damage resistance of laser coatings," in Laser Induced Damage in Optical Materials, Glass, A.J., Guenther, A.H., eds., US GPO Washington, DC: NBS Special Publication 541, 1978, pp. 190-201.

31. Wang, V., Guiliano, C.R., Garcia, B., "Single and multilongitudinal mode damage in multilayer reflectors at 10.6 pm as a function of spot size and pulse duration," in Laser Induced Damage in Optical Materials, Glass, A.J., Guenther, A.H., eds., US GPO Washington, DC: NBS Special Publication 435, 1975, pp. 216-229.

32. DeShazer, L.G., Newnam, B.E., Leung, K.M.,"The role of coating defects in laser induced damage to dielectric thin films," in Laser Induced Damage in Optical Materials, Glass, A.J., Guenther, A.H., eds., US GPO Washington, DC: NBS Special Publication 387, 1973, pp. 114-123.

33. Gill, D.H., Newnam, B.E., McLead, J., "Use of non-quarter wave designs to increase the damage resistance of reflectors at 532 and 1064 nanometers," in Laser Induced Damage in Optical Materials, Glass, A.J., Guenther, A.H., eds., US GPO Washington, DC: NBS Special Publication 532, 1977, pp. 260-270.

34. Aleshin, I.V., Doman, A.A., Imas, Ya.A., "Optical breakdown of glass by a sharply focused laser beam," Sov. Tech. - Phys. Lett., Vol. 4, No. 7, pp. 348-349, 1978.

35. Tuckerman, D.B., Weisberg, A.H., IEEE Electron Dev. Letter. EDL-7, 1 (1986).

36. Liu, R., Cheung, K.P., Lai, W.W.-C., Heim, R., Proc. 6th Int. IEEE VLSI Multilevel Interconnection Conf., p. 329, Santa Clara, CA (1989).

37. Jukai, R., Sasaki, N., Nakano, M., MRS Symp. Proc. 74, 229 (1987). also published in IEEE Electron Dev. Lett. EDL-8, 76 (1987).

38. Baseman, R., Andreshak, J., Gupta, A., Ting, C-Y., in "Selected Topics in Electronic Materials," ed. by B.R. Appleton and D.K. Biegelsen, EA 18, 259 (Materials Research Society, Pittsburgh 1989).

39. Broadbent, E.K., Ritz, K.N., Maillot, P., Ong, E., Proc. 6th Int. IEEE VLSI Multilevel Interconnection Conf., p. 336, Santa Clara, CA (1989).

40. Baseman, R.J., J. Vac. Sci. Technol. B-8, 84 (1990).

41. Rainer, F, Lowdermilk, W.H., Milam, D., Hart, T.T., Lichtenstein, T.L., Carniglia, C.K., Scandium oxide coatings for high-power UV laser applications, Applied Optics 21(20): 3685-3688; 1982 October 15.

42. Newnam, B.E., Gill, D.H., Ultraviolet damage resistance of laser coatings. Nat. Bur. Stand. (U.S.) Spec. Publ. 541: 190-201; 1978 December.

43. Turner, A.F., Ruby laser damage thresholds in evaporated thin films and multilayer coatings. Nat. Bur. Stand. (U.S) Spec. Publ. 356: 119-123; 1971 November.

Chapter 8

Deep-UV Microlithography

8.1 Introduction

Deep-UV lithography represents a highly specialized application of excimer lasers (1), a primary light source for imaging VLSI reticle patterns onto photoresist-coated silicon wafers for advanced IC fabrication. The excimer laser is integrated in a high resolution optical projection aligner, the wafer stepper. Deep-UV steppers are used in the volume production of high density memory, microprocessors, and high density bipolar and MOS logic circuits (2). Figure 8.1 shows a schematic of a deep-UV mask pattern generator used to produce IC patterns, and Figure 8.2 shows a deep-UV wafer stepper which projects the mask patterns onto silicon wafers to produce IC chips.

In this chapter, we will address the evolution of deep-UV lithography resulting in the sub-half-micron wafer stepper (3-5); technical requirements for deep-UV lithographic technology applied to manufacturing (6); excimer laser stepper functions and specifications; line-narrowed excimer laser design, spectral narrowing, and stabilization(7,8); production use of a deep-UV wafer stepper; and laser maintenance and utilities. Photoresists for deep-UV lithography (9-11) are covered in Chapter 4, "Material Research."

8.2 Evolution of Deep-UV Lithography

Integrated circuit fabrication, the primary application for deep-UV lithography, has undergone remarkable progress since the invention of the transistor in 1945. Figure 8.3 shows a single transistor as it appeared in actual size in 1945. A high IC density chip containing approximately 1 billion transis-

CHAPTER 8. DEEP-UV MICROLITHOGRAPHY

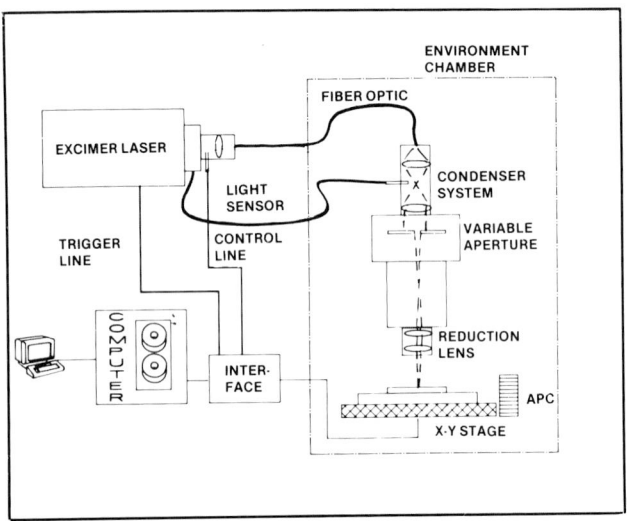

Figure 8.1: Schematic of a Deep-UV Mask Pattern Generator

tors (a Gigabit memory chip) can fit into the same area as this early single transistor device.

The density of IC devices has doubled every 2.5 years, driven by the ability to reduce the size of structures produced with optical lithography. Deep-UV optical lithography is the logical extension of longer-wavelength (i.e., 365 nm or the I-line of the mercury spectrum) optical imaging technology. Deep-UV imaging exploits the shorter wavelengths of deep-UV radiation, permitting smaller structures to be created on silicon wafers (12).

The useful result of microminaturization made possible by optical lithography is lower cost per bit of information storage space in computers. As the information handling capacity of computers doubles biyearly, the price of the computer is nearly halved in the same time frame. The DRAM (dynamic random access memory) chip represents the driving technology for smallness. DRAM memory chip evolution has been possible by the use of optical lithography (13).

8.2. EVOLUTION OF DEEP-UV LITHOGRAPHY

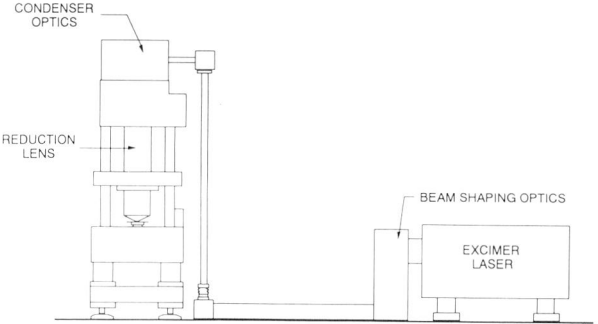

Figure 8.2: Schematic of a Deep-UV Wafer Stepper

Figure 8.3: Early Solid State Transistor

8.3 Deep-UV Technology: Resolution

Achieving higher resolution is fundamental to increasing the density of IC devices. The Rayleigh equation, which describes the resolution parameters in optical lithography is:

$$R = \frac{k1 \times lambda}{N.A.} \quad (8.1)$$

where R = resolution
K1 = k factor, an adjustable constant
lambda = exposing wavelength
N.A. = numerical aperture of the lens system

Resolution in optical lithography is more complex than the Rayleigh criteria indicates, and therefore must also be discussed along with focus depth. Total lithographic performace includes both resolution and depth of focus. The formula used to calculate focal depth is:

$$\frac{K_2 \times Lambda}{(N.A.)^2} \quad (8.2)$$

where DoF = depth of focus
and K_2 = an adjustable constant

8.4 Resolution: The "k" parameter

The adjustable constant "k" in the resolution equation includes any techniques or modifications to the lithography materials or equipment which impact resolution (15). K_1 is expressed as a number ranging from 0.8 down to 0.6, depending on the level of technology being used. For example, using 436-nm (G-line) exposing wavelength technology, 0.8 is a typical k parameter; with 365-nm (I-line) optical lithography, 0.6-0.7 is more typical as a "k" value; for deep-UV technology at 248 and 193 nm, the "k" parameter may be reduced further to 0.5 or below, depending of the resist system and other process variables, such as top-surface imaging technology, which will permit very low "k" values.

8.4. RESOLUTION: THE "K" PARAMETER

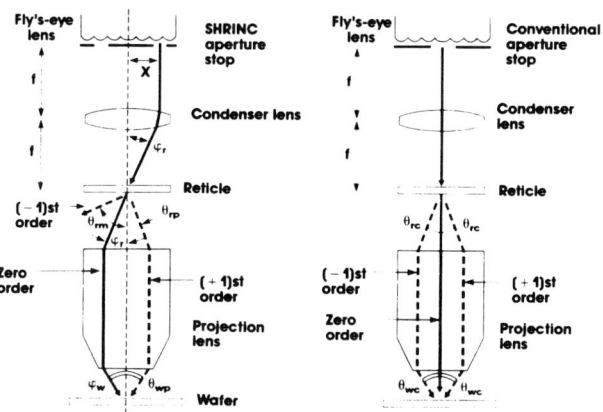

Figure 8.4: Conventional Illumination vs Annular Illumination

Other examples of technology variables that may impact the "k" parameter include phase shift masks, high contrast photoresists, and annular illumination schemes on wafer steppers. Phase shifting in general is a practical technique used to improve resolution, and shifting the phase of the light can occur at the mask plane or in the optics illuminator.

Annular illumination is a powerful and relatively simple method for improving both resolution and depth of focus. Annular illumination is a method of apertured exposure in a wafer stepper that changes the optical pathway and incident landing angle of incoming light. In conventional optical systems, zero and higher order diffracted beams interfere at the wafer plane to form an image. Since they have different incident angles, wavefront aberrations caused by path differences result in loss of resolution and limited focal depth.

Using strategically placed apertures, zero and (+1)st order beams enter the wafer plane at the same angle; all incident light enters at the same angle, permitting horizontal and vertical structures to be imaged identically. Figure 8.4 shows a comparison of conventional illumination and annular illumination (14).

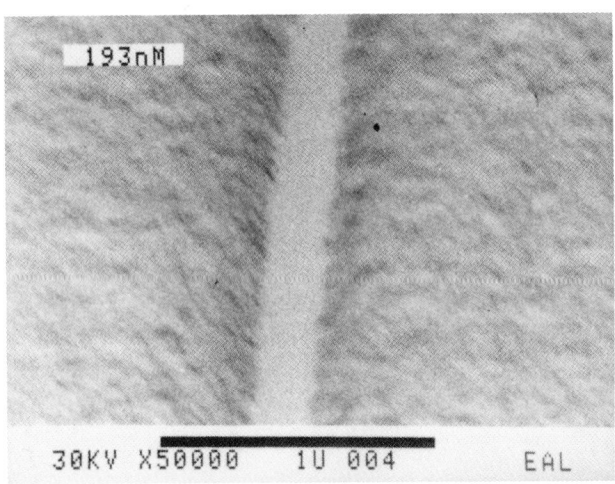

Figure 8.5: Submicron Pattern Made with 193 nm Imaging

8.5 Exposing Wavelength

The exposing wavelength in optical lithography has gradually decreased from 436 nm (G-line) to 365 nm (I-Line) to 248 nm (KrF) and 193 nm (ArF). Excimer laser lithography will make use of short wavelength light down to 193 nm, and potentially to 157 nm, the output wavelength of a fluorine laser. Beyond 157 nm lies the regime of X-ray lithography. Figure 8.5 shows a 5000x SEM of a resist patterned with a single laser pulse from an early (1988) 193-nm imaging tool (courtesy of Excimer Laser Systems).

8.5.1 Numerical Aperture

Higher numerical apertures in microlithographic lenses have added resolution to optical lithography. Routine manufacturing of lenses with 0.5-0.6 N.A. has permitted submicron optical imaging to be a volume manufacturing process for advanced IC devices. Field sizes of these lenses have also increased from 10 to over 20 mm^2 (16).

8.6 Deep-UV Laser Wafer Stepper

I-line lithography has extended the resolution capability of optical lithography down to 0.5 μm (16). I-line imaging with enhancements can achieve resolution down to 0.4 μm and slightly below. Deep-UV lithography will provide resolution in production from 0.4 to 0.2 μm and below. The highest resolution exposure tool (deep-UV stepper) will be used for the most critical dimensions and a lower resolution tool (i.e., I-line stepper used for noncritical features).

The excimer laser-based wafer stepper is used to manufacture advanced IC devices requiring geometries in the 0.4- to 0.2-μm range. Design rules with these small dimensions require the short UV wavelengths that only UV energy sources can provide, such as 248-nm krypton fluoride and 193-nm argon fluoride excimer lasers. Excimer laser-based wafer steppers will be used in conjunction with the 36-nm I-line steppers used for volume IC manufacturing.

Special imaging enhancements, such as annular illumination, phase shift masks, and higher N.A. lenses, extend image resolution with all exposure technologies. However, the process latitude required for volume manufacturing begins to run out for I-line lithography at 0.4-0.35 μm, and this is the regime where the deep-UV lithography technology takes over. Decreasing the exposing wavelength is the simplest and most cost effective way to provide increased resolution.

Deep-UV wafer steppers are manufactured in one of two categories: a) 5X reduction system based on an all-refractive lens system (14) and b) a catadioptric scanning system (15) at 4X demagnification. The reduction lens in the all-refractive system images the reticle pattern directly on the wafer, while the catadioptric reflective-refractive system images the reticle pattern on the wafer by scanning a slit exposure source across the reticle. The optical schematics of these two main types of lenses are shown in Figure 8.6.

The refractive lens-based stepper requires a highly spectrally narrowed laser; spectral narrowing is reduced or eliminated for the mirror-based stepper lenses since the effects of chromatic shift and resultant focal changes are less. The numerical aperture of the lens is related to the required bandwidth of the light source.

Fused silica lenses used in the all-refractive steppers have no chromatic correction capability. The higher the N.A., the greater the degree of spectral narrowing required. Typically, 0.5 N.A. lenses require narrowing to less than

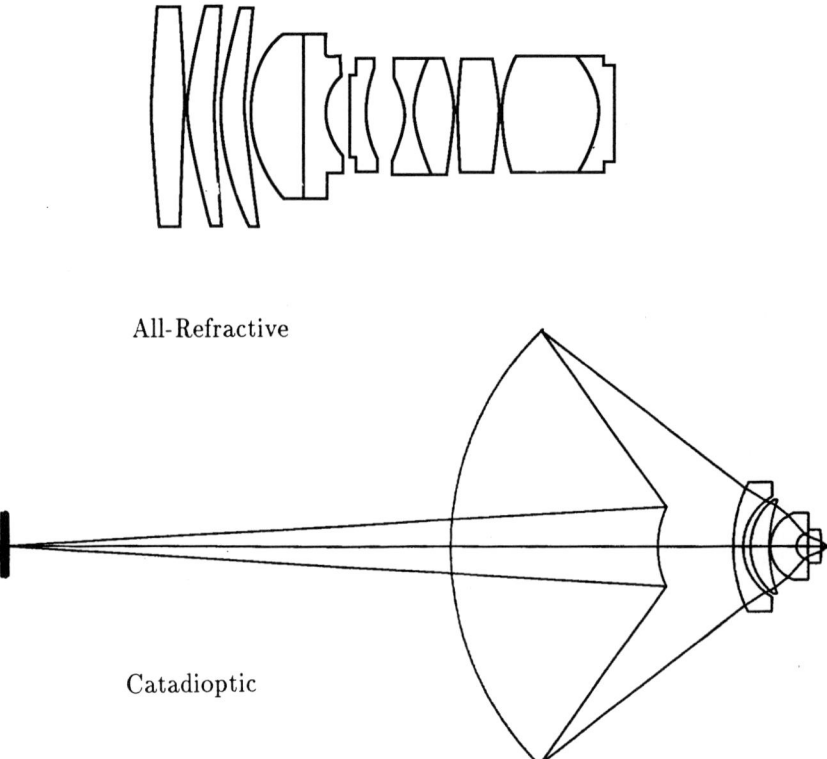

Figure 8.6: Optical Schematics of All-Refractive (a) and Catadioptic (b) Deep-UV Lenses

1 pm for maximum resolution.

Several key performance areas of the stepper relate directly to the laser light source. Several of these correlating parameters are:

Throughput: In volume IC manufacturing, wafer throughput of the stepper is a key measure of productivity. The output power of the laser is a key parameter in keeping wafer throughput high.

Resolution and Focus Depth: Resolution and focus depth are major imaging parameters for the stepper. The laser spectral bandwidth and spectral energy distribution relate most directly to resolution and focus depth. Wavelength stability also impacts resolution and focal depth.

Focal Plane Stability: The stepper must maintain constant focus over the entire imaging field. The relative wavelength stability of the laser over time is key in keeping the stepper focal plane stable.

Energy Dose Accuracy: Delivering a constant energy dose from one die site to another is necessary in lithography, so the laser must maintain a high degree of pulse-to-pulse stability.

Uniform Exposure: Highly uniform light is required at each die site to maintain critical dimension control in optical lithography. The laser beam profile and alignment are the main factors in keeping uniformity high. In addition, the stepper will incorporate beam homogenizers to mix or "uniformize" the beam prior to its entering the imaging lens. Laser beam divergence stability is also important. Optimized spatial coherence in the laser will also support exposure uniformity and reduce speckle.

8.7 Optical Materials for Deep-UV Imaging

Optics for deep-UV wavelengths must be of particularly high purity and be highly transmissive in the spectral region of interest. Conventional optical glasses used in lens making will not transmit usefully below 350 nm, and deep-UV lithography requires 248- and 193-nm wavelengths. Optical materials that provide high transmission of deep-UV radiation are quartz, lithium fluoride (LiF), magnesium fluoride (MgF_2), calcium fluoride (CaF_2), and fused silica (SiO_2).

The metal fluorides are crystalline in nature and have thermal expansion coefficients that are too high for advanced deep-UV lithographic lenses. Only fused silica can be manufactured with sufficient purity at sizes economical for large lenses; fused silica is thermally stable, noncrystalline, and highly transmissive at the deep-UV wavelengths of interest (248 and 193 nm). The

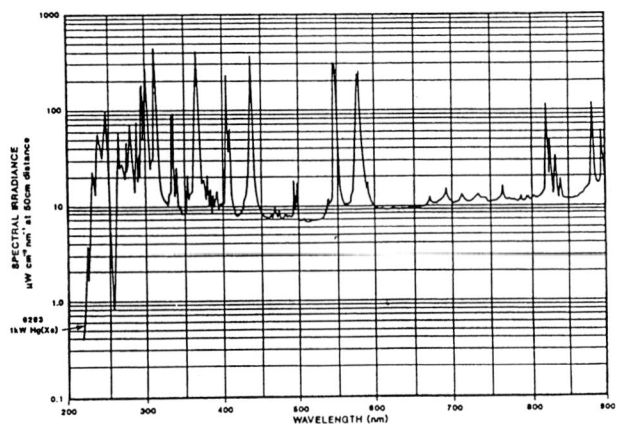

Figure 8.7: Mercury Lamp Spectrum

properties of high purity fused silica are given in Chapter 5, "UV Optics and Coatings."

High purity quartz and fused silica have a natural bandwidth of 300 pm and, with no chromatic correction, require that the laser bandwidth be optically narrowed in order to preserve the focal depth and resolution of the imaging optics.

8.8 Deep-UV Light Sources: Mercury Lamps

There are several sources of deep-UV radiation. The most commonly used are either incoherent lamp sources or coherent laser sources. The mercury arc lamp is widely used as a rich source of both visible and UV radiation, with a wide distribution of wavelengths and strong spectral lines in areas needed to expose photosensitive materials. A mercury lamp spectrum is shown in Figure 8.7.

Mercury lamps operate with input power in the 0.25- to 5.0- kW region. The spectral output is determined by the partial pressures of mercury and other dopant gases, such as xenon. They are configured in a quartz envelope at pressures up to 500 psi. A large portion of the input power of high and

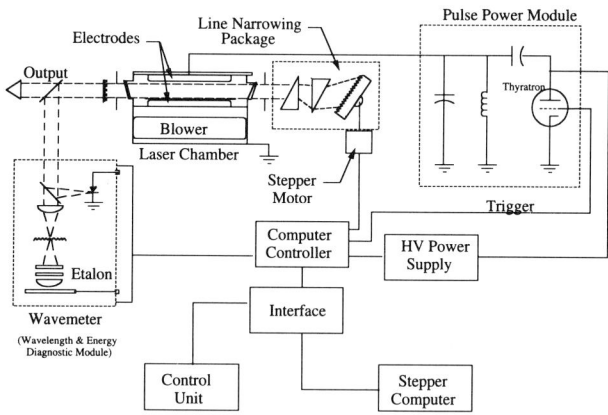

Figure 8.8: Lithography Laser Block Diagram: Key Subsystems

low pressure mercury lamps is converted into heat, so the deep-UV energy output is low. The output of these lamps in the 365- to 436-nm wavelength range has made them production light sources for optical lithography.

8.9 Line-Narrowed Excimer Laser Subsystems

The excimer laser is well suited as a source for advanced optical lithography, primarily because the output is spatially incoherent (coherence length of 100-200 μm). Too much coherence will produce "speckle," a wave-harmonic phenomenon that degrades resolution. Excimer lasers also possess relatively good optical efficiency (2% unnarrowed, 0.6% narrowed), can maintain constant output power with computer control, can be remotely installed, and have become cost competitive with other deep-UV sources.

The excimer laser for microlithography consists of several major subsystems. These are described below, along with their function and relationship to the other subsystems. The block diagram of the laser, shown in Figure 8.8, is taken from a Cymer line-narrowed laser used on major wafer steppers. The gas facilities are separate as shown in the figure.

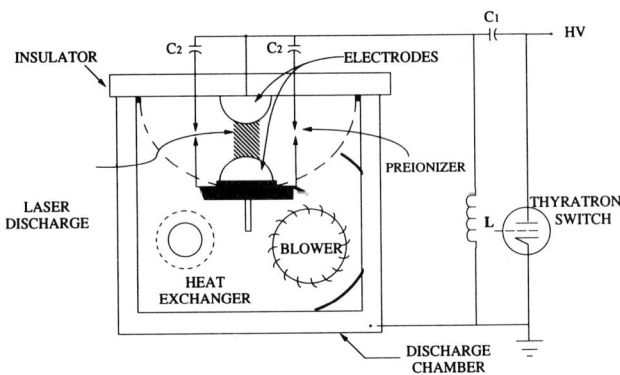

Figure 8.9: Schematic of Laser Discharge Chamber/Electrical System

8.9.1 Laser Head Module

The laser head module consists of the discharge chamber and the metal fluoride trapping system. The discharge chamber is broken down into the electrode structure, pre-ionizer, heat exchanger, blower, and windows.

8.9.2 Electrical Energy Transfer System

The electrical energy transfer subsystem consists of the pulse power module and thyratron support module. The pulse power module consists of the main capacitor, thyratron, and peaking capacitors. The thyratron consists of a trigger board, heater and reservoir, and a high voltage power supply. A schematic of the discharge chamber and electrical energy transfer system is shown in Figure 8.9.

The pulse power module and the laser discharge chamber are kept physically close together to minimize stray inductance and for rapid energy transfer into the discharge. The actual discharge area (inside chamber volume=50 x 1.5 x 0.5 cm) includes two electrodes with a separation of about 1.5 cm. The cathode is separated by an insulating ceramic substrate, and the anode is attached to the metal chamber.

8.9. LINE-NARROWED EXCIMER LASER SUBSYSTEMS

The pre-ionizer generates a UV corona through the use of high voltage surface currents; special structures are placed strategically on either side of the discharge region. Following each pulse, the discharge area must be cleared of the ambient gas, mainly because it contains metal particles that have ablated off the electrodes, as well as spent gas species, before the next pulse. This is accomplished by a blower which is magnetically coupled to an external drive source. Heat is removed from the laser by using a water-cooled heat exchanger inside the chamber.

The high voltage discharge causes ablation of metal particles from the electrodes. Metal particles are also generated by reaction of the gas with the chamber surface materials. These particles must be extracted continually from the chamber or they will coat the windows and degrade optical performance by reducing power and increasing bandwidth.

The laser described here uses a metal fluoride trapping system to remove the ablated electrode dust particles and other corrosion by-products from the chamber. Some lasers freeze out the particles by using a cryogenic processor through which the gas is continuously circulated.

The use of proprietary dielectric or inert materials inside the chamber greatly reduces this metal dust formation problem. It is critical that the insulating materials chosen are not classes of organics that will leech other contaminants (hydrocarbons, silicon, sulfur) in the presence of fluorine, high voltage, and ionic fluorine. For example, Teflon and Viton will leech out hydrocarbon to form CF_2, CF_4, COF_2, SiF_4, CO_2, HF_6, and SF_6.

These molecules will reduce power through photon absorption, deposit films on the laser windows, cause surface etching of the window, accelerate electrode wear, and reduce gas life. Materials for seals and electrical insulators must be chosen to be free of contaminants. Pure ceramics and high purity metals are good choices.

The pulse power module is made up of several components, including a thyratron switch, ceramic peaking capacitors, low inductance storage capacitor, thyratron trigger circuit, and a high voltage power supply. The thyratron is a hot cathode, deuterium-filled, high-voltage shorting switch, capable of switching high currents (15 kA) in a short time (30 nsec) at 400-500 Hz. The thyratron must fire only when the high voltage circuit is fully loaded, not sooner, since a prefire will cause a missing or low energy pulse. Typical performance is less than one prefire in 10^7 pulses, with an operating life of 2×10^7 pulses.

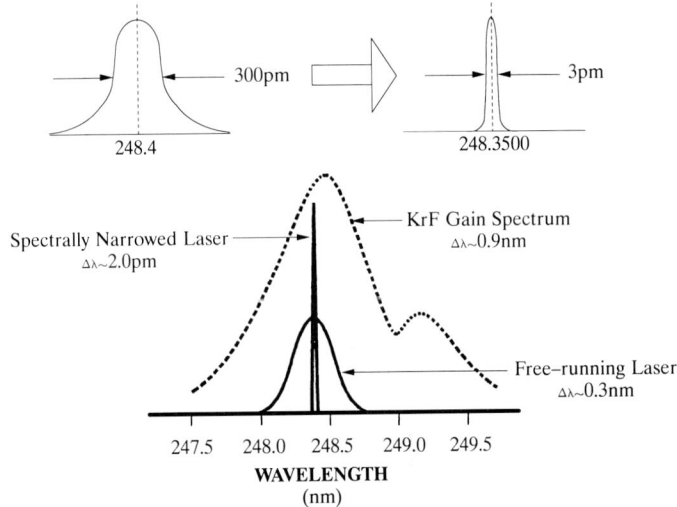

Figure 8.10: KrF Gain Profile and Required Spectral Reduction

8.9.3 Optical Resonator System

The optical resonator in a standard laser consists of a rear-mounted, maximum reflecting mirror and a front-mounted output coupler that is a partial reflector. A wavelength and energy diagnostic module are used to monitor energy, and the optics are held on a stable, unitized structure.

8.9.4 Spectral Narrowing

In a spectrally narrowed laser, a line-narrowing subsystem is added to the rear of the laser. This module reduces the optical bandwidth from the free-running width of about 300 pm down to 1-3 pm. Figure 8.10 shows the KrF gain profile and the narrowing required for laser stepper lenses (all-refractive). The final specification for spectral narrowing is determined by the N.A. of the imaging lens.

Lens systems that are mirror-based require very little to no spectral reduction since mirrors transmit all wavelengths to the same focal position. The addition of some refractive elements to mirror optical systems necessarily creates the need to compensate for the wavelength-induced focal changes. These catadioptric lens systems are desirable at shorter wavelengths where photon energy is high and all-refractive designs are subjected to photon dam-

8.9. LINE-NARROWED EXCIMER LASER SUBSYSTEMS

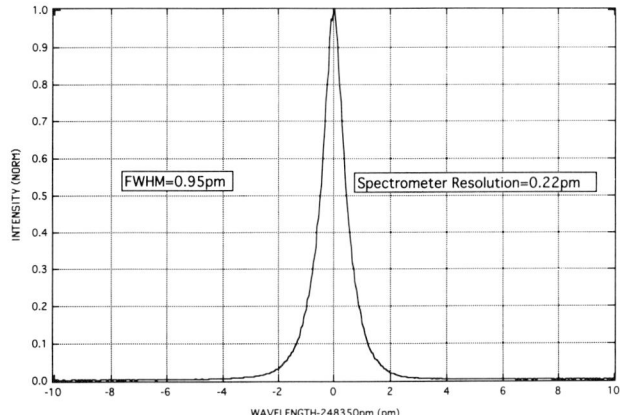

Figure 8.11: 248-nm Narrowed Excimer Laser Spectrum

age and very complex, multielement configurations to achieve resolution and near-distortion-free performance over large (30 x 30 mm) field sizes.

The principle of spectral narrowing relies on the introduction of wavelength dispersive optical elements in the resonator. Several types of dispersive elements are candidates for narrowing, including etalons, gratings, and prisms. Prisms do not provide sufficient dispersion, but are used to expand the beam onto a more dispersive element. Etalons are both efficient and highly dispersive elements, but are subject to optical damage and thermal distortion and drift.

Gratings are highly dispersive, damage resistant, and thermally stable. Gratings can be made with sufficient optical purity and flatness to deliver very high spectral quality. Since the gain medium acts as an amplifier, it is important that the quality of the grating be high. Further, a large area grating will provide reduced optical loading and thermal effects, which will increase the stable performance and life of the element in a line-narrowing system. Figure 8.11 shows the narrowed laser spectrum output on a high-resolution (0.22 pm) grating spectrometer. The undeconvolved bandwidth (full width at half maximum) is 0.95 pm.

In a line-narrowing module, the laser beam is first expanded with prisms onto the surface of the grating. The grating diffracts the incident light back

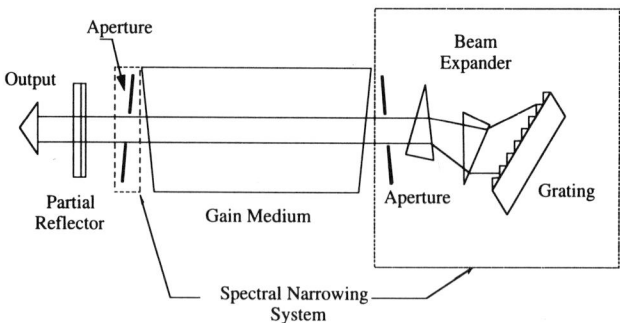

Figure 8.12: Line-Narrowing Schematic on a Lithography Excimer Laser

in the same direction as the incident beam (Littrow configuration) and into the gain medium. The final optical bandwidth is determined by several factors. Slit apertures at either end of the gain medium are used to select the spectral bandwidth. The grating parameters and the sample of light taken from the grating output further restrict the bandwidth; the number of optical round trips in the cavity will all figure in final bandwidth. Simple angular changes of the grating position are used to adjust the wavelength.

The need for increasing resolution in IC lithography has increased the need for higher N.A. lenses, which in turn has increased the need for narrower spectral bandwidth from excimer laser light sources. The optical materials used for line narrowing must also improve in quality. Examples include increased flatness and higher homogeneity of the optics since slight deviations in flatness or material purity can result in wavefront errors and subsequent spectral broadening. Small errors in the wavefront are magnified during lasing in the gain medium. Specifications for nearly defect-free optics are required.

Figure 8.12 shows a schematic of a line-narrowed excimer laser for lithography. This configuration uses a grating in Litrow configuration. The prisms are arranged to provide beam expansion, and the aperture is used for bandwidth control.

8.10 Coherence

The coherence properties of the laser are impacted by the laser cavity, laser optics, number of laser round trips, and other factors. A cavity with flat-flat optics (front and rear reflecting surfaces are flat) is more likely to produce speckle than a flat-concave cavity, where one optical surface will mix the modes. A flat-flat cavity keeps the coherence length constant. Longer pulses produce high coherence lengths. In an excimer laser, pulses are relatively short (17 nsec) with completely random wavefronts, eliminating or minimizing high spatial coherence. Longitudinal coherence gives rise to standing waves, but generally the delivery beam optics (diffusers, scanning mirrors, etc.) wash out high coherence.

8.11 Measurement of Wavelength and Bandwidth

The center wavelength of the laser needs to be stabilized in order to minimize changes in both focus and magnfication in the wafer stepper. The device used to measure a center wavelength is a wave meter. The wave meter is part of the line-narrowed laser assembly and therefore needs to be compact, yet highly accurate and precise. The wave meter must be accurate to +/-0.15 pm for a 1.5-pm bandwidth, and long-term drift must not exceed +/-0.15-pm. In addition, the precision of the wave meter, referenced to an atomic source, must also be specified to +/-0.15 pm. The overall operating range of the wave meter is 248.35+/-0.30 nm.

Wavelength is measured using a grating in combination with an etalon, shown in the wavemeter schematic in Figure 8.13.

The laser beam is sent through a slit aperture and onto the surface of the grating, which is housed in a pressurized chamber. The grating is used for coarse wavelength measurement. The beam is reflected off the grating spectrometer, off a mirror, and into the central region of a 1024 element silicon photodiode array.

A fine wavelength measurement is taken by sending the beam through a lens and diffuser, and into a second pressurized housing where the fine etalon generates a fringe pattern. The position of the coarse grating output and the diameter of the etalon fringe pattern, both shown in Figure 8.14, determine the wavelength of the laser output.

The calibration of absolute wavelength is made at the factory using a hollow cathode Ne-Fe lamp with an absorption peak at 248.3271 nm. It is important that the wave meter be both stable and reliable since it does not have its own internal absolute reference. Periodically, the wave meter can be sent to the factory to ensure that it has retained its preset reference wavelength.

268   CHAPTER 8. DEEP-UV MICROLITHOGRAPHY

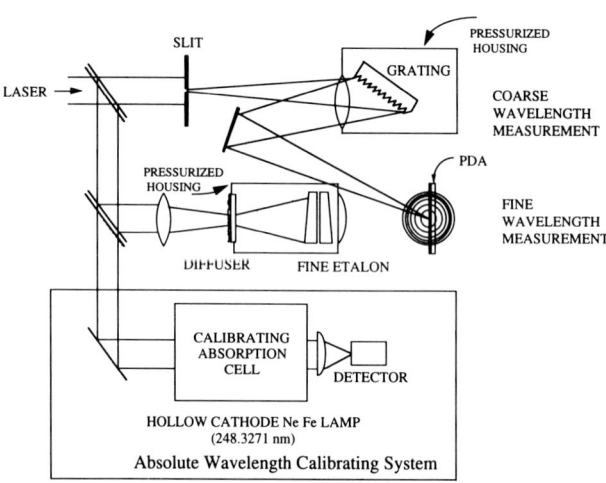

Figure 8.13: Schematic of a Wave Meter for a 248-nm Lithography Laser

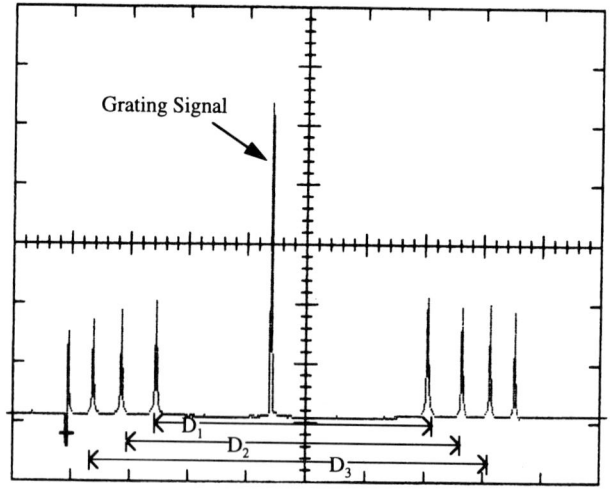

Figure 8.14: Wave Meter Output: Center Fringe=Coarse Grating Output; Outer Peaks=Fine Etalon Fringes

8.12. LASER PERFORMANCE 269

Etalons are widely used for wavelength stabilization due to their tolerance for changes in the input beam direction and their ability to control alignment errors in a simple mechanism. The free spectral range (FSR) of the etalon is 20 pm, and this fringe pattern is the same for all wavelengths which are separated by multiples of the FSR. The coarse grating measurement is used to prevent drift outside the specified FSR position. The fine etalon is then left to measure the wavelength within the preset FSR.

Typical specifications for the coarse grating are $+/-0.5$ pm; its long term accuracy is also $+/-0.5$ pm. The fine etalon resolution is $+/-0.1$ pm, with long-term accuracy of $+/-0.25$ pm. These figures are related to a laser narrowed to a specification of 1.5 pm. In taking measurements of the spectral bandwidth, the etalon resolution or bandwidth must be factored into the reading. For example, if the etalon bandwidth is 0.67 pm, the etalon fringe pattern measurement must be deconvolved by a factor of 0.67 pm. Further, there is some amount of energy in the "wings" of the spectral profile; spectral content in the wings must be measured separately on a high resolution spectrometer.

In summary, wavelength data from the wave meter are used to control laser wavelength by angular changes of illumination on the surface of the grating in a line-narrowing module. Laser bandwidth is fixed at the factory by the installation of a nonfield-tunable aperture.

8.12 Laser Performance

The main performance specifications for a lithography laser include the center wavelength, wavelength tuning range, repetition rate, pulse energy, power, spectral bandwidth (FWHM), band for 95% of the total energy, pulse-to-pulse energy variations, and long (6 months)- and short (8 hr)-term wavelength drift.

The requirements for wafer exposure throughput on the stepper determine the required power from the laser. A typical specification would be 500 Hz at 15 mJ/pulse, or 7.5 W of power. The photospeed of the deep-UV resist (30-60 mJ/cm^2) also determines the need for power from the laser. Spectral narrowing to a specified level is determined by the stepper lens resolution specification. Imaging at sub-quarter-micron resolution levels typically requires spectral narrowing below 1.25 pm.

The total integrated spectral energy output of the laser must be measured since this relates directly to resolution performance from the wafer stepper

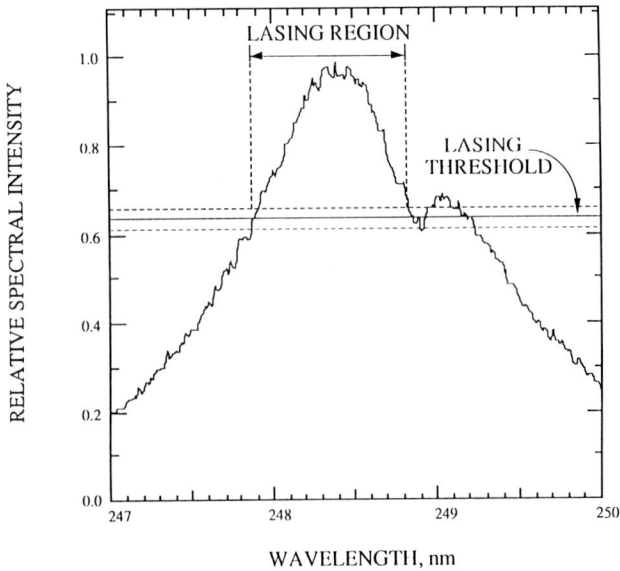

Figure 8.15: Spectral Intensity vs Wavelength

lens. Figure 8.15 shows the spectral energy integral measured on a high-resolution grating spectrometer. Note that a large percentage of the usable energy falls within the specified "spectral" boundary.

8.13 Laser Maintenance

The 248-nm excimer laser for microlithography is designed as a tool for the production floor, and therefore maintenance must be established on a regularly scheduled basis. The primary maintenance steps for this type of laser are:

Gas Exchange: Gas turnover is necessary to maintain a constant supply of fresh gas or a minimal level of fluorine atoms to support the energy demands of the laser. Gas exchange is specified in total shots per chamber fill or in time based on duty cycle or shot count.

A typical gas exchange specification is 20-30 million shots. The main symptom associated with a gas problem is deterioration of laser output. Laser chamber wear and optics are also causes of reduced laser output. Gas

8.13. LASER MAINTENANCE

problems include aging gas and low concentrations of fluorine species. Contamination builds up from the fluorine, reacting with chamber materials, causing production of CF_4, SiF_4, SF_6, and CF_2.

The continuous passivation of electrodes also causes the consumption of fluorine. Another source of contamination is from gas cylinders; if allowed to operate below 3000-4000 psi, cylinder wall material and cylinder bottom contamination can be introduced into the chamber. Plumbing leaks and pump oil backstreaming are other sources of contamination.

Laser gas lifetime can be improved through the proper choice of materials in the chamber, proper electrode materials, more frequent fluorine injection, and reduced operating voltage. An external gas processor can be added to freeze out and remove ablated electrode contamination and fluorine dust.

Fluorine Trap: The fluorine trap absorbs the effluent gas so that none is allowed to escape into the environment. This is a module that is periodically replaced. A site glass indicates by a color change when the trap is saturated. A typical fluorine trap can absorb about 250 laser chamber fills.

Window Cleaning: This is the most frequent scheduled maintenance step. Windows become dirty from particulates of ablated electrode material and "dust" generated in the chamber from reactions of fluorine with internal chamber materials. Windows are removed, cleaned with a solvent and tissue, and replaced. Typical cleaning time is 45 min, performed every 500 million shots.

Optics Replacement: At some point in the life of the laser, optics may become damaged by color centering (solarization) or etched by the gas medium. Damaged or etched optics are removed, repolished in some cases (if just etched), and reinstalled. Color center formation is internal (in the bulk of the optic) and permanently damages the optic.

Laser Chamber Replacement: The laser chamber eventually wears to the point where it must be rebuilt or replaced. Electrodes and high voltage components wear first, along with moving parts such as motors, bearings, fan parts, etc. Ideally, the chamber is replaced as a module. Modular design is highly desirable for a production tool, so that rapid replacement of modules keeps the laser stepper running productively (or available for running) 90-95% of the time. A typical chamber replacement will take place after 1 billion pulses.

Line-Narrowing Module: The life of a line-narrowing module depends on the quality of the original components, the frequency of cleaning and adjusting, and the energy density on the optical surfaces. In general, this module sees little wear, and expected lifetime is over 3 billion shots.

New technology and designs are being introduced regularly to extend the lifetime and reliability of all components in excimer lasers, especially the ones used with high duty cycles in manufacturing processes. Extended maintenance intervals and lifetimes will lower the usage cost and make excimer lasers more competitive with alternative energy sources.

The 248-nm line-narrowed excimer lasers for production use in IC manufacturing are widely accepted. Modular design to facilitate the rapid replacement of subsystems is widely used. Modules are repaired off-line while the laser stepper is kept operational.

8.14 Utilities

The main utilities required for installation of an excimer laser are laser gases, purging gases, cooling water, power, and exhaust. The gas mixture for 248-nm lithography lasers is 98% neon, 1.2% krypton, and the balance (0.1%) fluorine. Premixed gas eliminates handling high F_2 mixtures.

In addition to the laser gas, helium is used for purging the laser chamber and gas lines. Nitrogen is used to purge the optics and operate pneumatic valves. The nitrogen gas should not have any organic compound which dissociates in the presence of 193- or 248-nm radiation.

References

1. Pol, V., "Excimer Laser-Based Lithography: A Deep UV Wafer Stepper," Proceedings of SPIE, Vol. 633, March 1986, p.6.

2. "Excimer Lithography: Optimal Solutions in Lens, Laser, and Resists," Nikkei Microdevices, May 1989.

3. Jain, K., "Advances in Excimer Laser Lithography," Proceedings of SPIE, Vol. 774, March 1987, p. 115.

4. Burggraaf, P., "Deep UV Lithography: Crossing the Half-Micron Threshold," Semiconducor International, August, 1989, p. 62.

5. Elliott, D. J., Ferranti, D., "Sub-Micron Lithography at 248nm and 193nm Excimer Laser Wavelengths," Proceedings of SPIE, Vol. 922, 1988, p. 476-482.

8.14. UTILITIES

6. Elliott, D.J., Pennelli, C.P., Sengupta, U.K., "Recent Advances in an Excimer Laser Source for Microlithography," J. Vac. Sci. Technol, Vol. 9, No. 6, Nov/Dec, 1991.

7. Sengupta, U., "Excimer Lasers for Microlithography: Working Knowledge and Practical Considerations," Internal Publication of Cymer Laser Technologies, San Diego, CA. Nov. 1991.

8. Rhodes, C.K., Ed., "Topics in Applied Physics," Vol. 30, Excimer Lasers, Springer-Verlag, 1979.

9. Lin, B.J., "Deep UV Lithography," J.Vac.Sci.Technology, Vol. 12, No. 6, Nov/Dec, 1975.

10. Elliott, D.J.,"Resist Processing with Excimer Lasers," KTI Microelectronics Seminar Proceedings, Nov. 1987.

11. Escher, G., Telpot, G., Mohdro, R., "Advances in Deep UV Resist Processing," Proceedings of SPIE, Vol. 1925, 1993.

12. Jain, K., Wilson, C.G., Lin, B.J., "Ultrafast Deep UV Lithography with Excimer Lasers," IEEE Electron Device Letters, Vol. EDL-3, March 1982.

13. SEMATECH Internal Publication, "Deep UV Lithography Workshop," May 1994.

14. Elliott, D., "Pathway to the Gigabit Memory," Lasers and Optronics, Part 1, June 1993, Part 2, July 1993.

15. Shiroishi, N., et al, "Shrinc: A New Imaging Technique for 64 Mb DRAM," Microlithography World, July/Aug 1992.

16. Lyons, C., "I-line Lithography for Advanced I.C. Fabrication", Solid State Technology, April, 1992.

Glossary of Terms

Actinic, Absorbance The difference between the absorbance of unexposed and exposed photoresist. Actinic absorbance attempts to identify wavelengths where maximum absorbance change occurs, indicating areas of sensitizer and thus functional speed.

Airy Disc The control part (85% of total energy) of the diffracted image that is produced by an unobscured, circular aperture; mathematically, the fourier transform of the pupil shape.

Annular Illumination A method of using an aperture in the optical systems of a wafer stepper to allow zero and higher order diffracted beams to form identical incident angles at the wafer plane, thereby eliminating wavefront errors that degrade resolution when incident ray angles are different.

Aperture An on-axis, light-restricting mask or object in an optical system; a circular, square, or other (polygonal) shaped physical object that blocks radiation in an optical system from the object side. In telescopes, the area of the primary or first lens or mirror element that collects light and forms the image of a targeted object.

Aspherize To depart deliberately from spherosity in working an optical surface.

Bandwidth In optics the range of frequencies (or wavelengths) over which an optical system performs (or is specified to perform).

Catadioptric An optical system that combines refractive and reflective elements to achieve focal power, where much of the power of the system is derived from reflective surfaces, and refractive elements used for aberration correction, resulting in a relatively (compared to all refractive) low distortion optical system.

Contrast The difference between the maximum optical transmission density of the image and the optical transmission density of the nonimage areas.

Entrance Pupil The point on the lens optical axis where the chief ray would intersect the optical axis - if not redirected by the lens.

Fiducial Any mark used to control alignment of several layers (levels, planes) of a device. (Synonyms: keys, alignment marks, butterflies, registration marks)

FSR Abbreviation for "free spectral range." Frequency difference between adjacent resonances in an optical cavity.

8.14. UTILITIES

FWHM Abbreviation for "full width, half-maximum." A term describing the place where the laser beam properties, such as bandwidth, are measured, i.e., measuring the full width of the beam at the halfway point between "O" and maximum points.

Gamma A term normally applied to photographic films and generally determined as the tangent of the angle formed by the straight line portion of the D log E curve when extended to the log E axis. Gamma is obtained as a consequence of plotting the log of the exposure in millijoules per square centimeter for a specific wavelength of light or other radiation used for the purpose against the optical density. Generally, the higher the gamma, the higher the contrast of the image obtained.

Line Narrowing A projection imaging exposure tool used to print patterns on a 5X reticle in a layer of photoresist which is typically coated on a silicon wafer. Wafer steppers expose one die site at a time, then step over to the next die site, and repeat the exposure. The stop and repeat process continues until the entire wafer is exposed.

Microlithography The science of imaging micron and submicron structures onto silicon wafers and other substrates using a photomask with patterns of the image and a photoresist coating on the substrate onto which the mask pattern is formed.

Numerical Aperture The light gathering power of an optical system element, such as a refractive lens, optical fiber, or mirror system.

Optical Path Difference Difference between a perfect spherical wavefront (image centered) and a real wavefront.

Output Coupler A partial reflector mirror on the front end of a laser through which the beam exits (and also reflects back to the rear mirror).

Pellicle A mylar or nitrocellulose film stretched over a frame that is placed on the backside and frontside of masks and reticles to capture dust and other particles. The frame is far enough away from the work so as to keep particles falling on the pellicle membrane out of focus. This eliminates printed defects and improves device yields.

Phase-Shift Masks A chrome-on-quartz or surface-etched quartz with patterns designed by CAD and IC pattern data to shift the phase of the light in selected areas so as to permit better overall patterning fidelity (1:1 reproduction of all pattern information) and to compensate for proximity effects and other nonlinear geometry-related patterning effects.

Pixel A picture element; the smallest resolved area or object in a given image.

Raleigh Limit The maximum resolving power of an optical instrument according to Rayleigh Criterion.

Reciprocity Law Failure Failure of an energy-sensitive system to display a reciprocal function between exposure time and dose. Many resists possess a time dependence for completion of exposure reactions within the resist layer.

Resolution The smallest image that can be clearly discerned with the instrument and technique used.

Sharpness The visual impression of distinctness in a photographic reproduction, as the edge of an image; the subjective effect of the physical property acutance or resolution of the edge.

Spatial Coherence A relative optical condition or property of rays across the cross section of a laser beam; high spatial coherence is when most of the light has the same phase in space (temporal coherence describes the same property except in time space). Low spatial coherence is when there are many phases of rays across the beam cross section.

Speckle A pattern of diffracted light from a laser beam that appears as "graininess" in the light, the result of many diffracted beams.

Spherical Aberration An axial/imaging/aberration in a lens where light rays from various areas of the lens focus at different (closer or further) distances from the lens.

Thyratron A subsystem of an excimer laser that acts as a switch releasing a high voltage.

UV Corona A ring of low intensity UV light inside the discharge chamber of an excimer which is used to "seed" or initiate the laser discharge; permits the lasing ion to initiate sooner and at lower voltage levels reducing wear on the electrodes.

8.14. UTILITIES

Vacancy A missing ion in a lattice point.

Wafer A semiconductor substrate, sliced from a crystalline ingot of silicon or gallium arsenide or sapphire, and polished on one side to an optical finish. Wafers or "slices" are then cleaned, patterned with a resist layer, etched, and doped. After several such operations and final metalization steps, the wafers are cut up into individual die.

Wafer Stepper A projection imaging exposure tool used to print patterns on a 5x reticle into a layer of photoresist which is typically coated on a silicon wafer. Wafer steppers expose one die site at a time, then step over to the next die site, and repeat the exposure. The step-and-repeat process continues until the entire wafer is exposed.

Wavelength The physical distance covered by one cycle of a sinusodial wave of electromagnetic radiation.

Reading Bibliography

1. Muller, H.G., Yuan, Y., Sheets, R.E., "Large Area Fine Line Patterning by Scanning Projection Lithography," IEEE Transactions on Components, Packaging and Manufacturing Technology, Part B: Advanced Packaging, Vol. 18, No. 1, p. 33-6, Feb. 1995.

2. Nobutoki, H., Kumada, T., Koezuka, H., "Molecular Design of Dissolution Inhibitor in Chemical Amplification Resist System by Molecular Orbital Method: Transparency and Reactivities with 'Protons,' " Japanese Journal of Applied Physics, Part 1 (Regular Papers and Short Notes), Vol. 34, No. 2A, p. 623-9, Feb. 1995.

3. Endert, H., Patzel, R., Powell, M., Rebhan, U., Basting, D., "New KrF and ArF Excimer Laser for Advanced Deep-UV Lithography," Microelectronic Engineering, Vol. 27, No. 1-4, p. 221-4, Feb. 1995.

4. Nalamasu, O., Reichmanis, E., Timko, A.G., Tarascon, R., "A Unified Approach to Resist Materials Design for the Advanced Lithographic Technologies," Microelectronic Engineering, Vol. 27, No. 1-4, p. 367-70, Feb. 1995.

5. "First International Symposium on Semiconductor Processing and Characterization with Lasers," Materials Science Forum, Vol. 173-174, 1995.

6. Perrone, M.R., Palma, C., Bagini, V., Piegari, A., Flori, D., Scaglione, S., "Theoretical and Experimental Determination of Single Round-Trip Beam Parameters in a Xe-Cl Laser," Applied Physics A (Materials Science Processing), Vol. A60, No. 4, p. 411-17, April 1995.

7. Foulon, F., Green, M., "Laser Projection-Patterned Etching of (100) GaAs by gaseous Hcl and Ch_3Cl," Applied Physics A (Materials Science Processing), Vol. A60, No. 4, p. 377-81, April 1995.

8. Rim, C.S., Cho, Y.M., Kong, H.J., Lee, S.S., "Four-Mirror Imaging System (Magnification +1/5) for ArF Excimer Laser Lithography," Optical and Quantum Electronics, Vol. 27. No. 5, p. 319-25, May 1995.

9. Elliott, D.J., "Sub-Micron Deep-UV Imaging with a Catadioptric Step-and-Repeat Exposure System," Proceedings of the SPIE – The International Journal for Optical Engineering, Vol. 1835, p. 52-61, 1993.

10. Haizing, Z., Qihua, Y., Yu, Y., "A New Ring Array Off-Axis Illumination System for Sub-Half-Micron Lithography," Proceedings of the SPIE – The International Journal for Optical Engineering, Vol. 1927, Part 1, p. 235-8, Vol. 1, 1993.

11. Haixing, Z., Yudong, Z., "Some Problems in 1:1 Broadband Excimer Laser Lithography," Proceedings of the SPIE – The International Journal for Optical Engineering, Vol. 1674, Part 2, p. 701-6, 1992.

12. Johnson, D.W., Hartney, M.A., "Surface Imaging Resist for 193 nm Lithography," Japanese Journal of Applied Physics, Part 1 (Regular Papers and Short Notes), Vol. 31, No. 12B, p. 4321-6, Dec. 1992.

13. Bolle, M., Lazare, S., "Characterization of submicrometer periodic structures produced on polymer surfaces with low-fluence ultraviolet laser radiation," Journal of Applied Physics, Vol. 73, No. 7, p. 3516-24, 1993

14. Tomo, Y., Kasuga, T., Saito, M., Someya, A., Tsumori, T., "0.35 um rule device pattern fabrication using high absorption-type novolac photoresist in single layer deep ultraviolet lithography: surface image transfer for contact hole fabrication," Journal of Vacuum Science and Technology B, Vol. 10, No. 6, 1993.

8.14. UTILITIES

15. Wu, E.S., Strickler, J.H., Harrell, W.R., Webb, W.W., "Two-photon lithography for microelectronic application," SPIE, Vol. 1674, Pt. 2, p. 776-82, 1993.

16. Bolle, M., Lazare, S., LeBlanc, M., Wilmes, A., "Submicron periodic structures produced on polymer surfaces with polarized excimer laser ultraviolet radiation," Applied Physics Letters, Vol. 60, No. 6, p. 674-6, 1992.

17. "Development of a laboratory extreme-ultraviolet lithography tool," SPIE, Vol 2194, p. 95-105, 1994.

18. Kunz, R.R., Allen, R.D., Hinsberg, W.D., Wallraff, G.M., "Acid-catalyzed single-layer resists for ArF lithography," SPIE, Vol. 1925, p. 167-75, 1994.

19. Hattori, T., Schlegel, L., Imai, A., Hayashi, N., Ueno, T., "Chemical amplification positive deep ultraviolet resist by means of partially tetrahydropyranyl-protected," Optical Engineering, Vol. 32, No. 10, p. 2368-75, 1994.

20. Kunz, R.R., Allen, R.D., Hinsberg, W.D., Wallraff, G.M., "Acid-catalyzed single-layer resists for ArF lithography," Optical Engineering, Vol. 32, No. 10, p. 2363-7, 1994.

21. Proyer, S., Stangl, E., Schwab, P., Bauerie, D., Simon, P., Jordan, C., "Patterning of YBCO films by excimer-laser ablation," Applied Physics A: Solids and Surfaces, Vol. 58, No. 5, p. 471-74, 1994.

22. Tichenor, D.A., Kubiak, G.D., Malinowski, M.E., Stulen, R.H., Haney, S.J., Berger, K.W., Nissen, R.P., Wilkerson, G.A., Paul, p.H., Birtola, S.R., Jin, p.S., Arling, R.W., Ray-Chaudhuri, A.K., Sweatt, W.C., Chow, W.W., Bjorkholm, J.E., Freeman, R.R., Himel, M.D., MacDowell, A.A., Tennant, D.M., Fetter, L.A., Wood, O.R., Waskiewicz, W.K., White, D.L., Windt, D.L., Jewell, T.E., "Development of a laboratory extreme-ultraviolet lithography," Sandia National Labs, Livermore, CA 1994.

23. Imai, A., Asai, N., Ueno, T., Hasegawa, N., Tanaka, T., Terasawa, T., Okazaki, S., "0.13 um pattern delineation using KrF excimer laser light," Japanese Journal of Applied Physics, Part 1, Vol. 33, No. 12B, p. 6816-22, 1995.

24. Higashiki, T., "The estimation and improvement of overlay error budgets for an excimer laser aligner: compensation for illumination optical system errors," Journal of Japan Society of Precision Engineering, Vol. 60, No. 5, p. 662-6, 1994.

25. Sasago, M., Katsuyama, A., Yamashita, K., Endo, M., Nomura, N., Tani, Y., "Chemically amplified positive resist for KrF excimer laser lithography," National Technical Report, Vol. 40, No. 1, p. 97-105, 1994.

26. Nakano, K., Maeda, K., Iwasa, S., Yano, J., Ogura Y., Hasegawa, E., "Transparent photoacid generator (ALS) for ArF excimer laser lithography and chemically amplified resist," SPIE, Vol. 2195, p. 194-204, 1994.

27. Matsuo, T., Yamashita, K., Endo, M., Hashimoto, K., Koizumi, T., Katsuyama, A., Sasgo, M., Nomura, N., "New technologies of KrF excimer laser lithography system in 0.25 micron complex circuit patterns," Symposium on VLSI Technology, Digest of Technical Papers, p. 157, 145-6, 1994.

28. Boettiger, U., Fischer, T., Grassmann, A., Moritz, H., "Aerial image analysis of quarter micron patterns on a 0.5 NA excimer stepper," SPIE, Vol. 2197, p. 402-11, 1994.

29. Smith, B.W., Turgut, S., "Phase-shift mask isues for 193 nm lithography," SPIE, Vol. 2197, p. 201-10, 1994.

30. Nakajima, M., Yoshioka, N., Miyazaki, J., Kusunose, H., Hosono, K., Morimoto, H., Wakamiya, W., Murayama, K., Watakabe, Y., Tsukamoto, K., "Attenuated phase-shifting mask with a single-layer absorptive shifter of CrO, CrON, MoSiO, and MoSiON film," SPIE, Vol. 2197, p. 111-21, 1994.

31. Tounai, K., Hashimoto, S., Shiraki, S., Kasama, K., "Optimization of modified illumination for 0.25 im resist patterning," SPIE, Vol. 2197, p. 31-41, 1994.

32. Sugihara, T., Mori, S., Fukushima, T., Takagi, J., "Novel annular illumination using controlled phase and transparency," Symposium on VLSI Technology, Digest of Technical Papers, p. 168, 95-6, 1994.

33. Ohfuji, T., Aizaki, N., "0.15 mu m ArF excimer laser lithography using top surface imaging with high contrast silylation agent B (DMA) DS," Symposium on VLSI Technology. Digest of Technical Papers, p. 168, 93-4, 1994.

34. Endo, M., Hashimoto, K., Yamashita, K., Katsuyama, A., Matsuo, T., Tani, Y., Sasgo, M., Nomura, N., "Challenges in excimer laser lithography of 256M DRAM and beyond," International Electron Devices Meeting, Technical Digest, p. 1022, 45-8, 1992.

35. Ago, T., Nakagawa, H., "Development for deep-UV pellicles," SPIE, Vol. 2087, p. 186-98, 1994.

36. Ho-Young, K., Cheol-Hong, K., Joong-Hyum, L., Woo-Sung, H., Young-Bum, K., "High performance lithography with advanced modified illumination," IEICE Transactions on Electronics, Vol. E77-C, No. 3, p. 432-7, 1994.

37. Tounai, K., Kasama, K., "Optimization of optical parameters in KrF excimer laser lithography for quarter-micron lines pattern," IEICE Transactions on Electronics, Vol. E77-C, No. 3, p. 425-31, 1994.

38. Sasago, M., Matsuo, T., Yamashita, K., Endo, M., Matsuoka, K., Koizumi, T., Katsuyama, A., Momura, N., "New technologies of KrF excimer laser lithography systems in 0.25 micron complex circuit patterns," IEICE Transactions on Electronics, Vol. E 77-C, No. 3, p. 416-24, 1994.

39. Ohfuji, T., Nalamasu, O., Stone, D.R., "Simulation analysis of chemically amplified positive resist for KrF lithography," NEC Research and Development, Vol. 35, No. 1, p. 7-22, 1994.

40. Viswanathan, R., Seeger, D., Bright, A., Bucelot, T., Pomerene, A., Petrillo, K., Blauner, P., Agnello, P., Warlaumont, J., Conway, J., Patel, D., "Fabrication of high performance 512Kb SRAMs in 0.25 um CMOS technology using X-ray lithography," Microelectronic Engineering (Netherlands), Vol. 23, No. 1-4, p. 247-52, 1994.

41. Wallraff, G.M., Allen, R.D., Hinsberg, W.D., Larson, C.F., Johnson, R.D., DiPietro, R., Breyta, G., Hacker, N., Kunz, R.R., "Single-layer chemically amplified photoresists for 193 nm lithography," Journal of Vacuum Science and Technology B, Vol. 11, No. 6, 1994.

42. Nakase, M., Inoue, S., Fujisawa, T., Kihara, N., Ushirogouchi, T., "Key technologies in lower submicron lithography: ultimate super resolution imaging system and chemically amplified resist using the self-solubility acceleration effect," Journal of Vacuum Science and Technology B, Vol. 11, No. 6, 1994.

43. Sendova, M., Hiraoka, H., "Sub-half-micron periodic structures on polymer surfaces with polarized laser irradiation," Japanese Journal of Applied Physics, Part 1, Vol. 32, No. 12B, p. 6182-4, 1994.

44. Iwabuchi, Y., Ushioda, J., Tanabe, H., Ogura, Y., Kishida, S., "Mono layer halftone phase-shifting mask for KrF excimer laser lithography," Japanese Journal of Applied Physics, Part 1, Vol. 32, No. 12B, p. 5900-2, 1994.

45. Fuller, G., "MMST lithography," Solid State Technology, Vol. 37, No. 1, p. 55-6, 58-9, 1994.

46. Brunsvold, W., Stewart, K., Jagannathan, P., Sooriyakumaran, R., Parrill, J., Muller, K.P., Sachdev, H., "Evaluation of deep UV bilayer resist for sub-half micron lithography," SPIE, Vol. 1925, p. 377-87, 1994.

47. Pavelcheck, E., Calabrese, G., Dudley, B., Jones, S., Freeman, P., Bohland, J., Sinta, R., "Process techniques for improving performance of positive tone silylation," SPIE, Vol. 1925, p. 264-9, 1994.

48. Kunz, R.R., Allen, R.D., Hinsberg, W.D., Wallraff, G.M., "Acid-catalyzed single-layer resists for ArF lithography," SPIE, Vol. 1925, p. 167-75, 1994.

49. Hattori, T., Schlegel, L., Imai, A., Hayashi, N., Ueno, T., "Chemical amplification of positive deep-UV resist using partially tetrahydropyranyl-protected polyvinylphenol," SPIE, Vol. 1925, p. 146-54, 1994.

50. Kajita, T., Ei-ichi, K., Ota, T., Miura, T., "Novel negative resists using thermally stable crosslinkers based on phenolic compounds, "SPIE, Vol. 1925, p. 133-44, 1994.

51. Oikawa, A., Santoh, N., Miyata, S., Hatakenaka, Y., Tanaka, H., Nakagawa, K., "Effect of using a resin coating on KrF chemically amplified positive resists," SPIE, Vol. 1925, p. 92-100, 1994.

52. Roschert, H., Eckes, C., Endo, H., Kinoshita, Y., Kudo, T., Masuda, S., Okazaki, H., Padmanaban, M., Przybilla, K.J., Spiess, W., Suehiro, N., Wengenroth, H., Pawlowski, G., "Delay time effects between exposure and post exposure bake in acetal-based deep UV photoresists," SPIE, Vol. 1925, p. 14-30, 1994.

53. Kuwabara, S., Uchida, N., Tojo, T., Higashiki, T., Yoshino, H., "An excimer laser stepper with through-the-lens alignment," Journal of the Japan Society of Precision Engineering, Vol. 59, No. 8, p. 1257-62, 1994.

54. Zou Haixing, Zhang, Y., "Study of excimer laser direct projective lithography," SPIE, Vol. 1927, Pt. 2, p. 926-9, 1994.

55. Smith, B.W., Gower, M.C., Westcott, M., Fuller, L.F., "A 193 nm deep-UV lithography system using a line-narrowed ArF excimer laser," SPIE, Vol. 1927, Pt. 2, p. 914-25, 1994.

56. Kim, D.H., Choi, B.Y., Jung, K.R., Jun, C.H., Jang, W.I., Kim, Y.T., Lee, J.H., Park, H.O., "Design and development of a prototype excimer laser based stepper," SPIE, Vol. 1927, Pt. 2, p. 892-8, 1994.

57. Wittekoek, S., van den Brink, M., Poppelaars, G., Reuhman-Huisken, M., Grassmann, A., Boettiger, U., "Wide field deep UV wafer stepper for 0.35 um production," SPIE, Vol. 1927, Pt. 2, p. 582-94, 1994.

58. Arugu, D.O., Green, K., Nunan, P., Terbeek, M., Crank, S.E., Lam Ta, Capsuto, E., Sethi, S., "0.35 micron excimer DUV photolithography process," SPIE, Vol. 1927, Pt. 1, p. 287-97, 1994.

59. Dijkstra, H.J., Juffermans, C.A.W., "Optimization of anti-reflection layers for deep UV lithography," SPIE, Vol. 1927, Pt. 1, p. 275-86, Vol. 1, 1994.

60. Kubota, S., Kumada, T., Koezuka, H., Hanawa, T., Morimoto, H., "Photoresist materials for excimer-laser lithography," Mitsubishi Denki Giho, Vol. 67, No. 5, p. 19-22, 1993.

61. Ogawa, T., Kimura, M., Gotyo, T., Tomo, Y., Tsumori, T., "Practical resolution enhancement effect by new complete antireflective layer in KrF excimer laser lithography," SPIE, Vol. 1927, Pt. 1, p. 2632-74, Vol. 1, 1994.

62. Zou Haizing, Yan Qihua, Yan Yu, "A new ring array off-axis illumination system for sub-half-micron lithography," SPIE, Vol. 1927, Pt. 1, p. 235-8, Vol. 1, 1994.

63. Yen, A., Partlo, W.N., Palmer, S.R., Hanratty, M.A., Tipton, M.C., "Quarter-micron lithography using a deep UV stepper with modified illumination," SPIE, Vol. 1927, Pt. 1, p. 158-66, Vol. 1, 1994.

64. Partlo, W.N., Tompkins, P.J., Dewa, P.G., Michaloski, P.F., "Depth of focus and resolution enhancement for i-line and deep UV lithography using annular illumination," SPIE, Vol. 1927, Pt. 1, p. 137-57, Vol. 1, 1994.

65. Sasago, M., Endo, M., Tani, Y., Kobayashi, S., Koizumi, T., Matsuo, T., Yamashita, K., Nomura, N., "Quarter micron KrF excimer laser lithography," IEICE Transactions on Electronics, Vol. E76-C, No. 4, p. 582-7, 1993.

66. Bolle, M., Lazare, S., "Large scale excimer laser production of submicron periodic structures on polymer surfaces," Applied Surface Science, (Netherlands), Vol. 69, No. 1-4, p. 31-7, 1993.

67. Heydel, R. Matz, R., Gopel, W., "Maskless excimer laser induced projection patterning of InP in Cl/sub2/ etch gas," Applied Surface Science (Netherlands), Vol. 69, No. 1-4, p. 38-45, 1993.

68. Miyaji, A., Suzuki, K., Tanimoto, A., "Excimer lithography for VLSI," Optical and Quantum Electronics, Vol. 25, No. 5, p. 297-310, 1993.

69. Yen, A., Partlo, W.N., McCleary, R.W., Tipton, M.C., "0.25 um lithography using a deep UV stepper with annular illumination," Microelectronic Engineering, Vol. 21, No. 1-4, p. 37-42, 1993.

70. Grassmann, A., Ittner, G., Wittekoek, S., "Production-worthy deep UV stepper for 0.35 micron lithography," Microelectronic Engineering, Vol. 21, No. 1-4, p. 25-8, 1993.

71. Nomura, N., Yamashita, K., Endo, M., Sasgo, M., "Lithography beyond 64 Mb," Microelectronic Engineering (Netherlands), Vol. 21, No. 1-4, p. 3-10, 1993.

8.14. UTILITIES

72. Kunz, R.R., Horn, M.W., "Dry-developed organosilicon resists for 193 nm excimer laser lithography," Polymer Engineering and Science, Vol. 32, No. 21, p. 1595-9, 1993.

73. Preil, M.E., Arnold, W.H., "Aerial image formation with KrF excimer laser stepper," Polymer Engineering and Science, Vol. 32, No. 21, p. 1583-8, 1993.

74. Leuschner, R., Borndorfer, H., Kuhn, E., Sebald, M., Sezi, R., Byer, M., Nolscher, C., "Progress in deep UV resists using CARL technology," Polymer Engineering and Science, Vol. 32, No. 21, p. 1558-64, 1993.

75. Nakase, M., "Recent progress in KrF excimer laser lithography," IEICE Transactions on Electronics, Vol. E76-C, No. 1, p. 26-31, 1993.

76. Matsuo, S., Komatsu, K., Takeuchi, Y., Tamechika, E., Mimura, Y., Harada, K., "High resolution optical lithography system using oblique incidence illumination," International Electron Devices Meeting, 1991.

77. Kunz, R.R., Horn, M.W., Wallraff, G.M., Bianconi, p.A., Miller, R.D., Goodman, R.W., Smith, D.A., Eshelman, J.R., Ginsberg, E.J., "Surface-imaged silicon polymers for 193 nm excimer laser lithography," Japanese Journal of Applied Physics, Part 1, Vol. 31, No. 12B, p. 4327-31, 1993.

78. Johnson, D.W., Hartney, M.A., "Surface imaging resists for 193 nm lithography," Japanese Journal of Applied Physics, Part 1, Vol. 31, No. 12B, p. 4321-6, 1993.

79. Tomo, Y., Kasuga, T., Saito, M., Someya, A., Tsumori, T., "0.35 mu m rule device pattern fabrication using high absorption-type novolac photoresist in single layer deep ultraviolet lithography: surface image transfer for contact hole fabrication," Journal of Vacuum Science and Technology B, Vol. 10, No. 6, 1993.

80. Kunz, R.R., Horn, M.W., Bianconi, P.A., Smith, D.A., Eshelman, J.R., "Wet-developed bilayer resists for 193 nm excimer laser lithography," Journal of Vacuum Science and Technology B, Vol. 10, No. 6, p. 2554-9, 1993.

Chapter 9

Micromachining

9.1 Introduction

The micromachining of components from sheets of metal or slices of silicon is used to produce a variety of 3D microcomponents. Microsensors and microactuators are examples of products that can be combined with integrated circuits to perform various external functions, such as regulating the flow of gas in an automobile or controlling the temperature and humidity in a large building. Millions of micromachined devices are used worldwide annually in automobiles and in many other mechanical and electronic systems (1).

The future integration of complex micromachined parts with equally complex microprocessor and other IC devices can create very cost-effective microelectromechanical systems (MEMS) for many applications in our society. The design, development, and commercialization of cost effective micromachined devices for high volume industrial and consumer applications will be critical in making MEMS technology realize its full potential.

The primary requirement for micromachined devices is that they can be manufactured in volume with good cost effectiveness. Advancements in process technology have been made which now permit this to occur. Technical requirements for micromachined include:

- low stress
- low stiction
- high shock resistance
- high thermal stability (hot and cold)

- internal physical strength
- good wear resistance
- and high corrosion and environmental resistance

These must also be realized to make micromachining technology and MEMS viable.

Devices produced by micromachining are made with quasi-planar processes using additive and subtractive techniques, similar in approach to the processes used to fabricate integrated circuits. Some of the same metal deposition techniques, etching equipment, and process steps for IC fabrication are used to micromachine materials such as silicon, PZT (ferromagnetic lead zirconate/lead titanate), silicon nitride, ceramic, and copper. The geometries of micromachined components are typically 10x larger than IC geometries; the most critical MEMS dimensions are typically 1 μm, while critical dimensions for ICs are about 0.1 μm (2).

The use of ultraviolet light for producing small structures by direct ablation is based largely on the high quantum energy per photon and the other special properties of UV light discussed in Chapter 1, "Ultraviolet Light." UV light at much lower fluence levels is also used for its short wavelength and corresponding high resolution properties in the lithographic processes which are part of most micromachining methods.

In this chapter the basic methods are reviewed for forming structures using UV light for ablation and imaging processes. UV micromachining and microforming processes are presented, along with details of the actual parameters used in various applications to produce working devices.

The main processes covered are traditional micromachining by direct UV ablation, patterning by masked ablation, mid-UV and deep-UV lithography to produce MEMS, and 3D microforming using UV light and imaging processes. Each of these techniques has characteristics suited for certain applications. Details of the various applications are given at the end of the chapter. A brief description of the basic principles and mechanisms for microstructuring and micromachining is given in the next section.

9.2 Micromachining Processes

Several processes can be used to create micromachined components. In this section the basic approaches are described in summary form, then covered

9.2. MICROMACHINING PROCESSES

in detail later in the chapter. The four methods described are: 1) Print and Etch Chemical Micromachining; 2) Direct Ablation; 3) 3D Microforming; and 4) Microelectromechanical Systems processes.

Print and Etch Chemical Micromachining: Early micromachining processes were based on photolithography (originally mercury vapor sources). This process used copper or metal substrates which were imaged with a photoresist on one or both sides and then etched through the thickness of the substrate. This is called "bulk" micromachining since the entire thickness of the substrate is etched through, leaving a solid piece of metal behind.

"Surface" micromachining differs from "bulk" processes in that only the surface of the substrate is processed through multiple lithographic, etching, and deposition steps. In both surface and bulk micromaching the result is a large number of components formed on a single sheet of material at a low cost per part. The main problem with this process is that it is restricted to easy-to-etch metals which are easily etched.

Surface micromachining is used today mainly to make inexpensive parts, such as Christmas ornaments, jewelry, and other decorative items. Brass, copper, and glass are typical materials used.

Direct Ablation: Direct ablation is a process where high fluence UV energy is used to directly etch materials, without photoresist, by projection exposure. The raw laser beam from conventional excimer lasers restricts the area to be imaged at one time, typically 1-2 inches square. Larger and more powerful (1 kW) industrialized UV lasers have made it possible to directly ablate areas up to 4-6 inches square, making it suitable for manufacturing applications in niche markets. An example is the direct ablation of polymers for multichip module package manufacturing (3). Figure 9.1 shows an example of this application, where a 248-nm excimer laser is used for direct ablation through an aperture.

3D Microforming: 3D microforming is a process where the thick photoresist images are used to contain the electroplating of metal on a substrate. After plating, metal inside the walls of the resist image is stripped, and the metallic plated components are peeled off the plating base. 3D microforming does not use any subtractive etching, and the resolution limit of parts that can be made is determined by the lithographic process.

Microelectromechanical Systems: The MEMS processing involves the use of additive and substractive techniques to create microcomponents of various materials, including silicon and metals. MEMS technology is being used to fabricate a wide variety of sensors, actuators, and components that can interface with IC devices to create a "smart" system. An example of a

Figure 9.1: Ablated Windows in Photoresist for IC Packaging

component generated with the MEMS process is shown in Figure 9.2 (14).

9.3 Direct Ablation

The basic mechanism of micromachining by direct ablation is based on the absorption of UV energy by various substrate materials. Most metals, glasses, polymers, and other dielectrics used in micromachining absorb UV energy very intensely, each with a characteristic absorption coefficient.

The absorption versus wavelength characteristics vary as a function of the material used. Efficient processing is possible by selecting the UV wavelength that most optimally couples with the material to be imaged or ablated. The 351-nm (xenon fluoride excimer) and 308-nm (xenon chloride excimer) wavelengths are strongly absorbed by polymers, such as novolak photoresists, polyimide, and a variety of plastics.

Metals, ceramics, glasses, and metal silicides absorb strongly at or near 248 nm (krypton fluoride). The hardest materials to micromachine, such as quartz, diamond, and dense or high grade aluminum oxide (ceramic), are best processed at 193 nm, where the higher photon energy is advantageous during the ablation reaction.

9.3. DIRECT ABLATION

Figure 9.2: Laser-Machined Gear for Micromotor

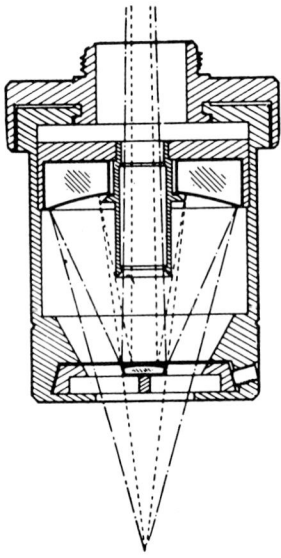

Figure 9.3: Two-mirror Schwarzchild Lens

9.3.1 Ablation Thresholds

Miror and catadioptric lenses are sued to micromachine metals, plastics, and semiconductors. A typical all-mirror lens is shown in Figure 9.3. The two-mirror Schwarzchild-type lens is effective in delivering relatively high laser fluence for micromachining at all principal excimer wavelengths.

One example of this lens in use is shown in Figure 9.4, where a plastic pellet is being drilled by two 193-nm beams. The holes are nearly drilled through, and these were performed without moving the z-stage.

The ablation threshold of a given material is defined as the fluence value that will cause the phenomenon of ablation. Fluence is expressed in J/cm^2. Ablation occurs as UV energy is absorbed, causing material excitation, heating, expanding, and exploding away at a high velocity from the surface of the absorbing material. The material leaving the irradiated solid cools and condenses, and portions of by-products land back on the surface unless scavenged by a nozzle or other means (5).

The ablation threshold for a material will vary according to several parameters, most of which are laser beam related. These include laser reprate (Hz), wavelength, pulse energy and duration, fluence or energy density, am-

9.3. DIRECT ABLATION

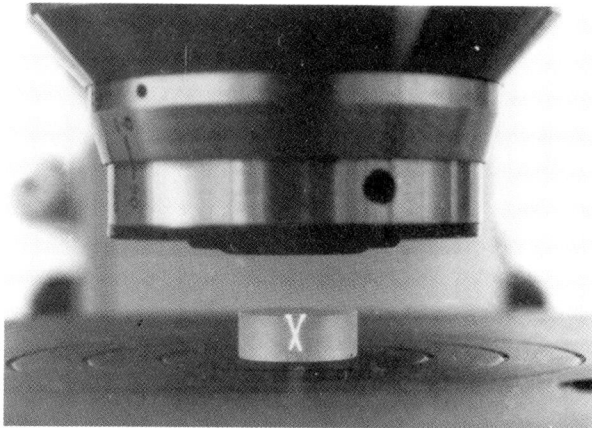

Figure 9.4: Plastic Machining at 193 nm (courtesy of IMS)

bient conditions at the material interface (pressure, temperature, gas, etc.), and the absorption properties of the substrate. The selection of conditions from these various parameters will determine the occurrence, depth, nature, and rate of the ablation reaction.

Figure 9.7 gives approximate ablation threshold values for a number of commonly used films in micromachining (6).

9.3.2 Direct Ablation Process

The process for direct ablation is a simple one-step exposure of a hose of UV energy. The UV beam can be either rastered back and forth across the substrate or sent to selected sites for ablation with or without a pattern recognition system.

Direct ablative machining is a relatively simple, one step process for direct structuring. Figure 9.5 shows where a focused excimer laser beam was used to ablate the plastic outside a copper lead frame, then, as shown in Figure 9.6, ablate the copper lead (8). The particular advantages of this method are simplicity of the process and high resolution. Focused spot imaging or the beam writing technique is good for producing prototype structures, but is not generally cost-effective for volume manufacturing. The disadvantage is that large areas are difficult to cover in a single shot or dose due to limited available UV energy.

294   CHAPTER 9. MICROMACHINING

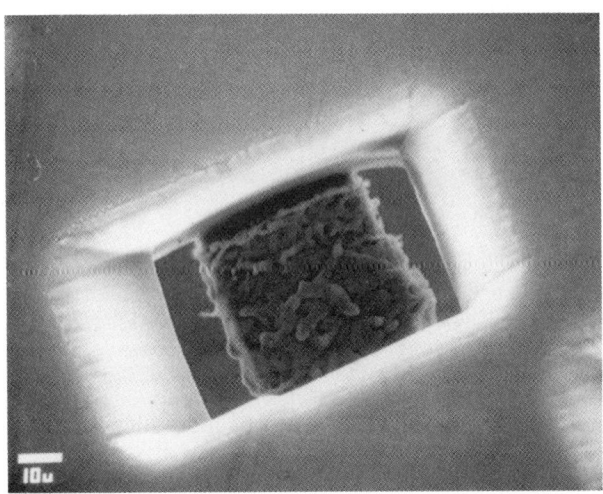

Figure 9.5: Ablation of Polymer over Copper Lead

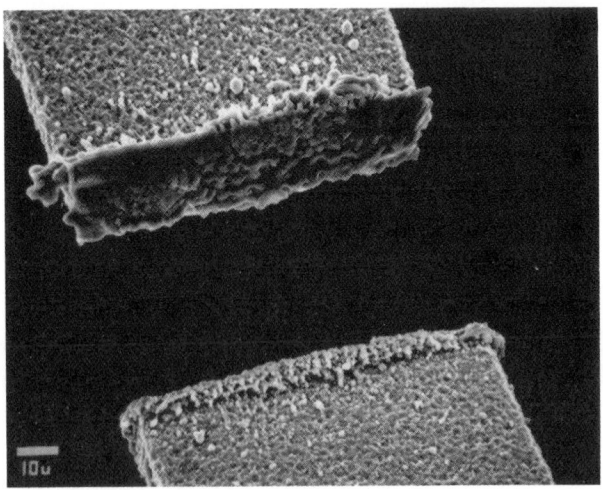

Figure 9.6: Ablation of Copper Lead

Nitrocellulose	> 20
Novolak-Based Photoresist	30
Polycarbonate	40
Polyimide	45
Polysilicon	100
Silicon Nitride	195
Spin-on-Glass	200
Silicon Dioxide	350
Glass, Metal Oxide	700-1200
Metal Films	4-8000

Figure 9.7: Ablation Thresholds of Various Materials: mJ/cm^2

In addition to ablation of low density materials, metals such as gold, copper and aluminum can be structured with UV light. Gold, copper, and aluminum are widely used in the electronics industry. These metals can be difficult to etch, requiring strong acids and highly baked photoresists. Direct ablation with 10-14 J/cm^2 fluence at 248 nm is sufficient to form patterns in many metals. The SEM photo in Figure 9.8 shows an aluminum oxide gear machined with a 248-nm laser at 50 J/cm^2. The gear diameter is 230 μm and requires approximately 10,000 pulses (11).

9.4 Micromachining Applications

9.4.1 Metal Machining

Submicron UV laser machining of small holes in plastics, semiconductors and metals is used to create many types of microdevices ranging from filters to micromotor parts. Many optical systems have been designed for hole drilling. The one shown in Figure 9.9 is a 248-nm excimer laser with a 200-mJ output and unstable resonator optics. The raw beam is attenuated to reduce the pulse energy to about 20 mJ. The iris aperture removes excess, nonuniform edges of the beam, and a Galilean beam expander is used to keep the beam divergence close to the diffraction limit. The 150-mm lens focal plant is imaged onto the substrate with a 10x reduction lens. The shaping aperture (iris no. 2) controls the spot size at the workpiece. The final lens

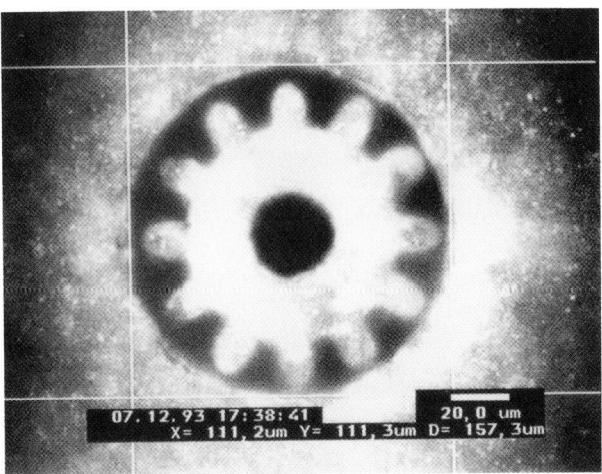

Figure 9.8: UV Laser Ablation of Aluminum Oxide Gears

is an all-quartz objective optimized at 200 nm.

The SEM of holes in Figure 9.10 are part of an electronic interconnect device, made of molybdenum. The melting point of molybdenum is 2617°, and the metal thickness is 15 μm. The input side where the laser first strikes the "moly" foil is on the right side of the figure, with a pronounced molten crater. The roundness of the hole at the exit side is due mainly to melting effects (12).

9.4.2 Ceramic Machining

Machining of aluminum oxides and zirconium oxides with excimer lasers is playing a role in several industries, notable electronics. A test (15) run was made comparing a Nd:YAG to a XeCL excimer laser. The SEM photos in Figure 9.11 compare a ZrO_2 cut with the Nd:YAG laser (left side) versus the excimer cut sample. The xenon chloride laser used a cutting speed of 15 mm/min with a 5-mm line focus and power of 23 W. The Nd:YAG laser required 160 W of average power and could only cut at a rate of 10 mm/min. Note the difference in the quality of the cut made.

The krypton fluoride laser machining of aluminum oxide has been well-documented in several published articles. It has been found that helium is a

9.4. MICROMACHINING APPLICATIONS

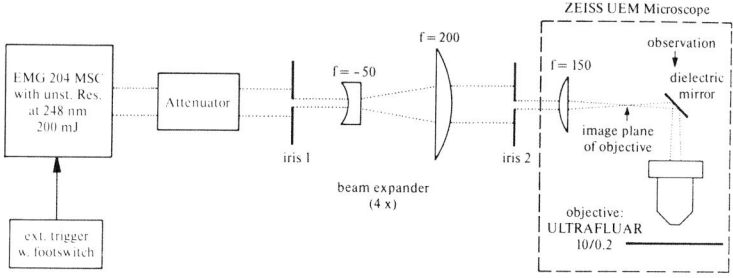

Figure 9.9: Optical System (MicroLas) for Submicron Machining

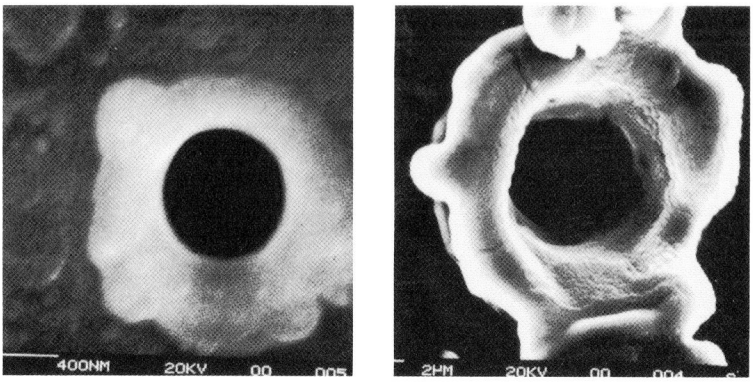

Figure 9.10: Molybdenum Foil Machined with 0.8 μm Holes (exit - left; entrance - right)

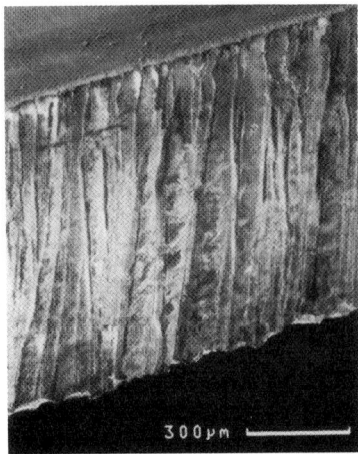 

Figure 9.11: Nd:YAG (Left) and XeCL (Right) Cutting of ZRO_2

good gas for flushing the workpiece to cool the sample, and simple focusing optics can be used. Figure 9.12 shows the number of pulses and ablation depth versus energy density used in machining a 0.65-μm sample of Al_2O_3 (16).

The same material was machined using several aperture diameters at a 38.8 demagnification. Data in Figure 9.13 show the close correlation between number of pulses and energy density (16).

9.4.3 Sapphire Machining

In another test comparing the excimer to Nd:YAG laser cutting, samples of sapphire were both micromachined. Figure 9.14 shows the excimer cuts (left) and cuts made with the Nd:YAG (right) at 1060 nm. The excimer laser wavelength was 193 nm (17).

9.4.4 MEMS Application: Atomic Force Microscope Probe

The probe tips for atomic force microscopes can be produced by patterning a resist layer on silicon dioxide and overetching the underlying silicon so as to produce considerable undercut, resulting in a probe tip with atomic-scale dimensions. Figure 9.15 shows the schematic layout of the silicon probe and support circuitry (18). A close-up of the probe is also shown.

9.4. MICROMACHINING APPLICATIONS

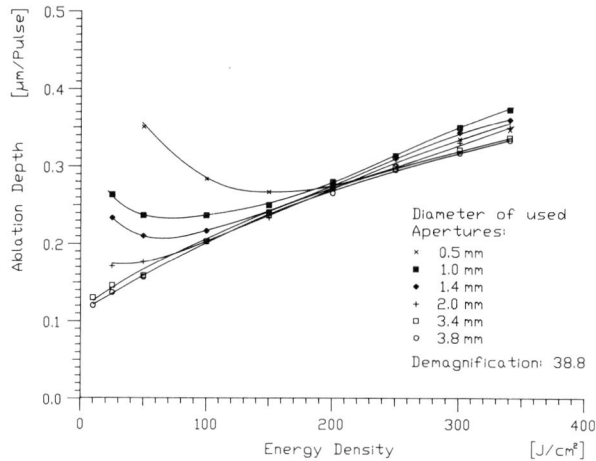

Figure 9.12: Aluminum Oxide Machining: at 248-nm Ablation Parameters

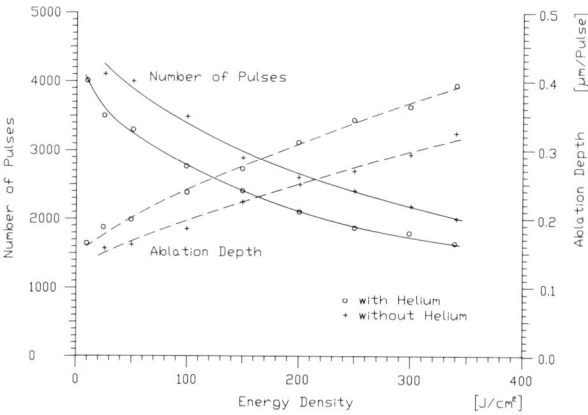

Figure 9.13: Aluminum Oxide Machining: at 248 nm Ablation Parameters

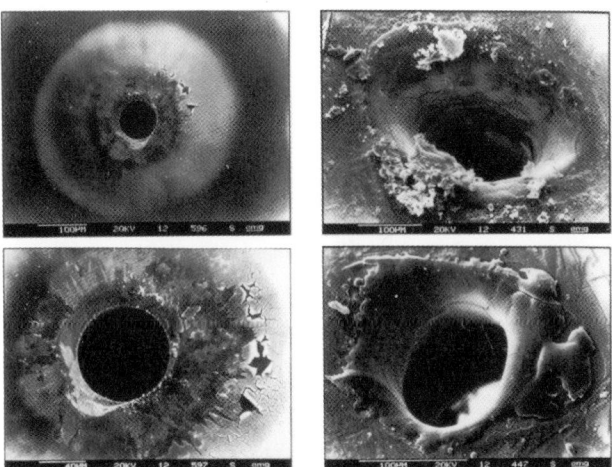

Figure 9.14: Sapphire Drilled with a 193-nm Excimer Laser (Left) and a Nd:YAG Laser at 1060 nm (Right)

9.5 Three-Dimensional UV Microforming

9.5.1 Introduction

A branch of micromachining is 3D microforming, a process used to fabricate three-dimensional components on a silicon wafer using ultraviolet proximity printing and electroplating processes. This is essentially an "additive" process since a photoresist image is first formed which is a negative impression of the final part. This "mold" is then electroplated to form the actual metal part, and the resist is dissolved away, leaving a solid component on a silicon substrate.

3D microforming is a lithography-intensive process and does not rely on etching to form the microsized components as is the case for traditional micromachining. In this section, we will review the process details and show examples of parts and applications

9.5.2 The Process

One of the main advantages of 3D microforming is that the parts can be manufactured using standard integrated circuit process equipment. In many IC fabs, lithography equipment may not be at capacity or obsolete UV align-

9.5. THREE-DIMENSIONAL UV MICROFORMING

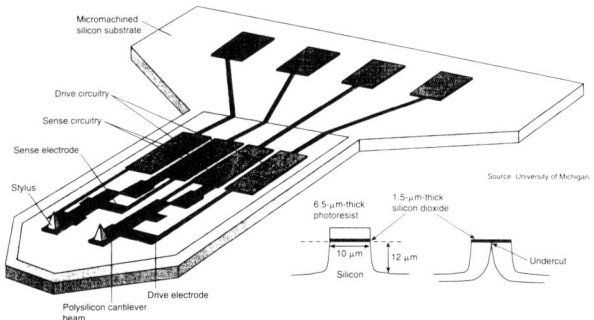

Overview of AFM Probe System

Close-up of Machined Tip

Figure 9.15: Atomic Force Microscope Probe Made with MEMS Technology

Figure 9.16: Stators and Wiring System of Electrostatic Micromotor

ers, such as proximity printers, can be reactivated for volume production of 3D-formed parts. Typical parts which can be manufactured with standard IC process equipment include rotors, microcoils, and magnetically driven valves.

A thick resist imaging process combined with standard electroplating are the two main processes needed to make parts with the 3D microforming technology. The stators and three-dimensional wiring system of an electrostatic motor are shown in Figure 9.16, fabricated with the process cited above.

The main process steps are described in detail as follows:

Resist Coating: The first step in a typical 3D microforming process is the application of a thick positive-tone photoresist layer. The substrate surface must be properly cleaned prior to the application of resist to assure good adhesion. Resists must be highly viscous to permit the application of a 50- to 60-μm-thick layer in a single step (9). The spin coater bowl and chuck area is covered to maintain a solvent atmosphere and to keep the resist surface from drying and sealing. Special spin coaters with rotating covers are also used for this process.

Softbaking: To prevent the entrapment of a solvent in such a thick layer, a multistage baking procedure is used, permitting complete desolvation and preventing the surface from drying first, thereby trapping the solvent inside the thick resist film. This type of softbaking (ramped with time and temperature) is commonly used for all thick photoresist processes.

Exposure and Development: The resist can be exposed by contact printing or proximity printing. Proximity printing is preferred since it does not cause degradation of the photomask as is the case in contact printing. In

9.5. THREE-DIMENSIONAL UV MICROFORMING

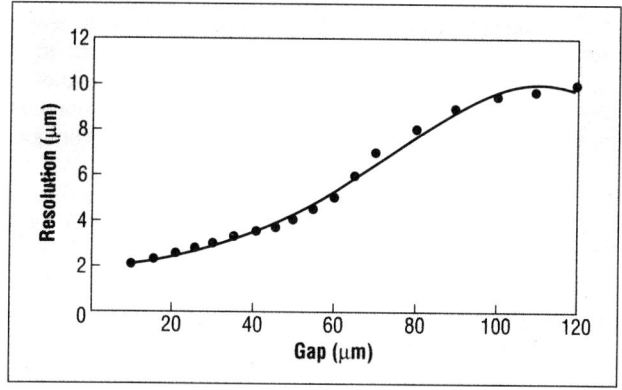

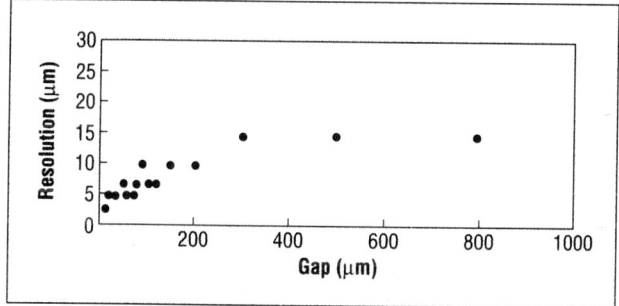

Figure 9.17: Resolution vs Mask-Wafer Gap

proximity printing, mask-to-wafer gaps of 10-250 μm can be used, depending on the mask resolution and the nature of the structures to be imaged. Figure 9.17 shows the trade-off between exposure gap and resolution.

Photoresist coatings over 100 μm thick may require two exposures at different depths since the depth of focus is less than 100 μm. The exposure tool is programmed for a dual-focus-expose step, permitting the lower half of the coating to be exposed first, followed by the upper half. Exposure to produce steep, 80° sidewall angles in the photoresist is necessary for components such as rotors, where the final angle on the metal must be nearly perpendicular.

Diffraction effects, such as standing wave patterns, often occur with highly collimated light and must be eliminated. A simple method to avoid

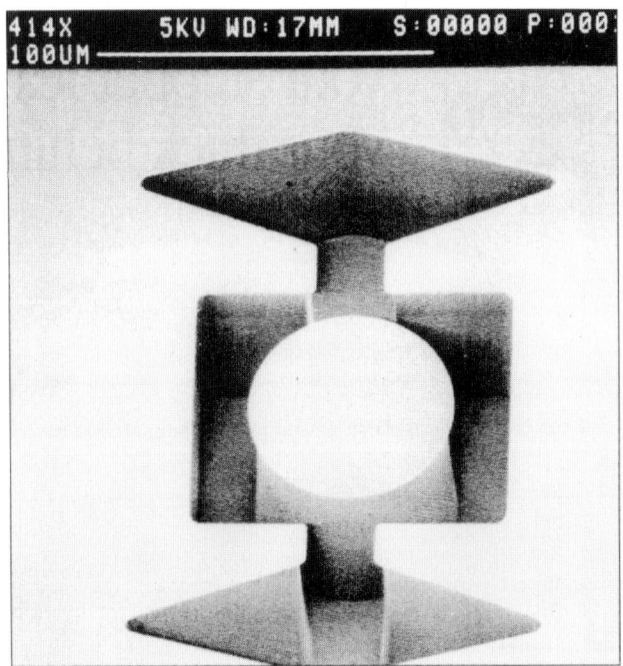

Figure 9.18: Photoresist with a 20-μm Feature/57 μm Thick

standing waves is to adjust the photoresist thickness (or the total optical thickness) so it becomes an odd quarter-multiple of the exposing wavelength. The use of an odd 1/4 multiple will avoid creating the reflection harmonic called standing waves wherein alternating exposure minima and maxima leave a saw-tooth pattern in the resist image sidewall. The optical thickness is the total thickness of material, usually just the photoresist, that light passes through before it is reflected back onto itself.

Developing is performed using simple immersion in an aqueous alkaline solution, followed by water rinsing and forced air drying. A SEM photo of imaged photoresist is shown in Figure 9.18.

Electroplating: Pulse and DC electroplating are used to deposit inside the walls of metal films, such as gold, nickel, copper, and nickel-iron alloys.

Gold plating is done in a standard sulfonic acid bath with a 10- to 30-μm/min deposition rate. Plating parameters such as current density and

9.5. THREE-DIMENSIONAL UV MICROFORMING

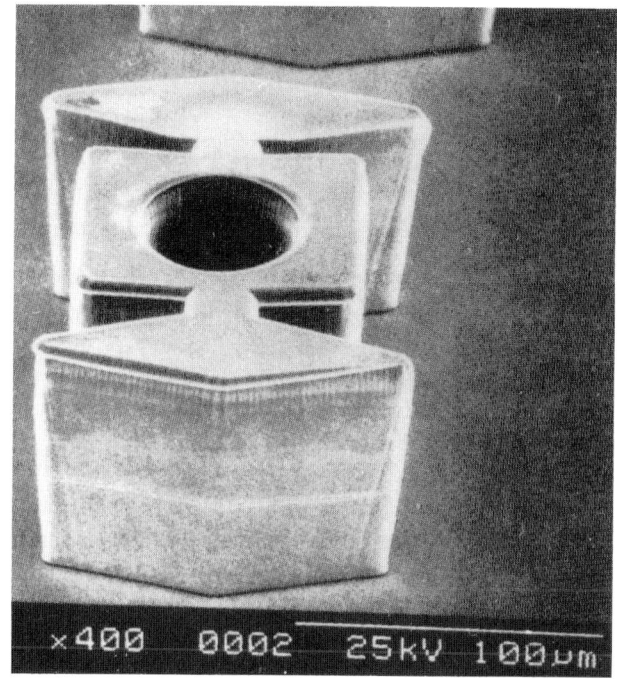

Figure 9.19: Microformed Nickel-Iron Rotor Element

solution flow are optimized for the most uniform metal deposits. Plating of nickel-iron alloys (85% nickel/15% iron) is also performed with commercially available plating baths. The parameters for this process are described in the open literature (4). Plating times, temperatures, current densities, and bath composition are controlled to optimize the metal deposits to meet the requirements of MEMS structures. MEMS technology is described in detail in the next section of this chapter.

Nickel-iron rotor elements, shown in Figure 9.19, are best made by depositing onto pure nickel bases. Gold plating bases prove to be the best metal bases for copper- and gold-plated components.

<u>Resist Removal</u>: The last step is resist removal, accomplished using a standard aqueous alkaline resist remover.

Solvents for stripping photoresist are not recommended due to their harmful effects on the environment, disorders caused to human skin, and absorption into the bloodstream where they have been linked to certain cancers. Resist strippers based on aqueous alkaline chemistry are preferred since they can be properly neutralized and waste treated.

9.6 Microelectromechanical Systems

9.6.1 Introduction

Microelectromechanical systems are miniature devices or clustered arrays of components which integrate both mechanical and electronic functions. MEMS are made using equipment similar (in some cases identical) to that used for fabricating silicon-integrated circuits. Many components for MEMS are devices fabricated on a silicon wafer substrate using the same etch, deposition, and lithography processes used in IC fabrication. Deep-UV lithography provides the high resolution imaging critical to MEMS fabrication.

Products made with this technology range from silicon pressure sensors to inertial sensors, fluid regulators and control valves, optical switches, and mass data storage devices. Emerging applications include blood pressure transducers, infusion pressure sensors, auto engine performance sensors, tire pressure sensors, air bag accelerometers, navigational gyros, motion control devices, micromechanical valves and regulators, fluid meters, optical access, rigid disk drive mechanical components, and archival high capacity/high rate data storage devices. Some of these devices are now produced in the millions, and will likely be produced in the billions as the technology matures and becomes increasingly cost-effective.

9.6.2 MEMS Processing

MEMS technology relies on "surface" micromachining, which is not limited by the aspect ratios of the etching processes, as is "bulk" micromachining. Surface micromachining does not penetrate the substrate, but instead uses alternating imaging, etching, and deposition processes to form mechanical structures. In some designs, integrated circuits are formed.

The MEMS process is relatively simple and extremely flexible. Surface micromachined parts can be smaller in size than bulk machined parts, but suffer from potential stiction and fragility compared to the bulkier and generally larger parts made by etching through the thickness of the substrate.

The process used to form mechanical structures (10) is simple: A cross-sectional view of a fabricated micromotor using this process is shown in

9.6. MICROELECTROMECHANICAL SYSTEMS

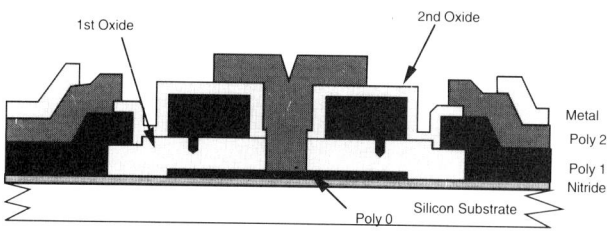

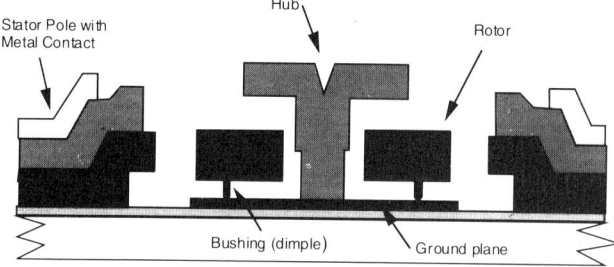

Figure 9.20: Micromotor Cross Section (a) as Fabricated and (b) after Release by Wet Etching Oxide

Figure 9.20. There are six successive layers of low pressure chemical vapor deposited (LPCVD) dielectrics (nitride, oxide, and polysilicon), followed by a final evaporated metal interconnect layer (10).

The substrate is a silicon wafer covered with a 0.5-μm layer of low stress silicon nitride, the isolation layer. The first of three layers of low stress polysilicon is then deposited and used as a ground plane layer. This layer is patterned and etched, followed by deposition of the first oxide layer, consisting of 2 μm of phosphorous doped glass (PSG). The second layer of "poly" is the first structural layer and is also 2 μm thick. This, like all the deposited layers, is patterned and etched. The next layer is the second oxide, used to connect the two poly layers and anchor the poly 2 layer. Figure 9.21 shows a micromotor fabricated with this process (10).

The third poly layer is 1.5 μm thick and is the second structural layer. After patterning, the metal layer is deposited, patterned, and etched just like the previous layers. This metal is called the "contact" layer since it makes electrical contact with the device and connection to the "outside world." The "as-fabricated" structure is then wet etched to remove all the oxide,

Figure 9.21: Micromotor Made with MEMS Technology

thereby "releasing" the microcomponents.

9.7 Failure Analysis Applications

UV laser machining into integrated circuit packages is a method suitable to analyze circuit failure mechanisms, especially since excimer laser ablation can be controlled to stop precisely at calculated depths. Figure 9.22 shows an opening ablated in the outer plastic IC package layer to expose the IC chip surface (8). Figure 9.23 shows a close-up of the aluminum metallization that can now be electrically probed and tested (8). Note the undamaged surface of the aluminum.

9.8 Wire Stripping and Fiber Taps

Ablation of the insulative cladding of electrical wire is an application used in high volume production. The wire in Figure 9.24 has been stripped by 308-nm ablation, and the underlying metal is unaffected (13). Figure 9.25 shows another micromachining application where 193-nm ablation was used to make a "T-joint" in an optical fiber (8).

9.8. WIRE STRIPPING AND FIBER TAPS

Figure 9.22: Excimer Ablation into an IC Device

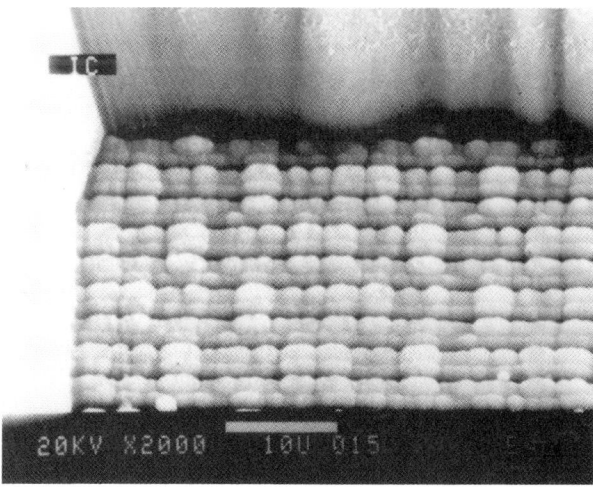

Figure 9.23: Close-Up of IC Active Device Following Ablation of Outer Package Layer

310   CHAPTER 9. MICROMACHINING

Figure 9.24: Close-Up of Wire with Insulation Cladding Ablated Off

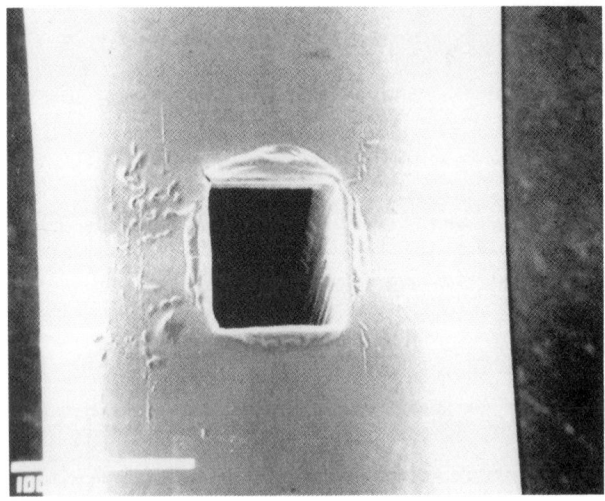

Figure 9.25: Optical Fiber with 193-nm Ablated Opening for a "T-Joint"

References

1. Core, T.A., Tsang, W.K., Sherman, S.J., Solid State Technology, p. 39, Oct, 1993.

2. Vintro, L., Solid State Technology, p. 57, April, 1995.

3. Elliott, D.J., Piwcyzk, B.P., "Electronic Materials Surface Processing with Excimer Lasers," Microelectronic Engineering, p. 435-444, North-Holland, Elsevier Science Publishers B.V., 0167-931, 1986.

4. Shukov,G.A., Smith, A., "Micromachining with Excimer Lasers," Lasers and Optronics, p. 75, 1988.

5. Srinivasan,R., "Ablation of Polymers and Biological Tisue by Ultraviolet Lasers," Science, Vol. 234, pp. 559-565, Oct. 31, 1986.

6. Technical Memorandum, Internal Publication of UVTech Systems Inc., 526 Boston Post Road, Wayland, MA., April, 1995.

7. Industrial Report Number 8, Internal Publication of Lambda Physik, November, 1994.

8. Leitz-Image Micro Systems, Application Photos, October, 1988.

9. Cullmann,E., Lochel, B., Engelmann, G., Reyerse, C., "3D Structures for Micro-System Technology Using Proximity Lithography," p. 93, March 1995, Solid State Technology.

10. MEMS Fact Sheet, Published by MCNC MEMS Technology Applications Center, 3021 Cornwallis Road, Research Triangle Park, North Carolina 27709-2889.

11. Lambda Physik "Highlights", No. 43, January, 1994.

12. Lambda Physik "Highlights", No. 2, December, 1986.

13. Lambda Physik "Highlights," No. 8, November, 1994.

14. Fraunhofer Institute for Silicon Technology, Internal Bulletin, Berlin, 1994.

15. Lambda Physik "Highlights" No. 17, June, 1989.

16. Lambda Physik "Highlights" No. 34, April, 1992.

17. Lambda Physik "Highlights" No. 18, August, 1989.

18. "Micromachines on the March", IEEE Spectrum, May, 1994.

Glossary of Terms

Ablation Threshold A specific energy input level, usually from a laser at which the bonding energy of an irradiated solid is overcome by input radiant energy and the solid decomposes by ablation.

Bulk Machining The deep etching of metals or other materials (semiconductors) to form microcomponents. A mask is typically used to define the boundaries of the part. Also called "chemical milling," since liquid etchings are often used.

LPCVD An acronym for "low pressure chemical vapor deposition." A process for laying down thin films of semiconductors using an evacuated reactor vessel with an anode and cathode

MEMS An acronym for microelectromechanical systems. MEMS devices are combinations of micromachined, microformed, and microelectronic devices (ICs) that are formed into a system to perform a specific function such as a sensor of actuator.

Optical Waveguide A physical structure that confines and directs radiant energy along its physical boundaries and parallel to its axis with minimal to zero loss.

Poly Short form for "polysilicon," a polycrystalline film layer used to fabricate "MEMS" structures.

PSG Phosphorous doped glass, an impurity-doped semiconductor film of silicon dioxide used to form "MEME" structures.

Surface Machining The etching of the top portion or thin surface layer of a material to define a shape or component pattern.

Three Dimensional Microforming A process involving the use of photolithography and electroforming to form 3D parts.

9.8. WIRE STRIPPING AND FIBER TAPS 313

Reading Bibliography

1. Bennett, T.D., Grigoropoulos, C.P., Krajnovich, D.J., "Near-Threshold Laser Sputtering of Gold," Journal of Applied Physics, Vol. 77, No. 2, p. 849-64, Jan. 1995.

2. Hibi, Y., Enombto, Y., Kikuchi, K., Shikata, N., Ogiso, H., "Excimer Laser Assisted Chemical Machining of SiC Ceramic," Applied Physics Letters, Vol. 66, No. 7, p. 817-18, Feb. 1995.

3. Arnold, J., Dasbach, U., Ehrfeld, W., Hesch, K., Lowe, H., "Combination of Excimer Laser Micromachining and Replication Processes Suited for Large Scale Production," Applied Surface Science, Vol. 86, p. 251-8, Feb. 1995.

4. Hrovatin, R., Mozina, J., "Optodynamic Aspect of a Pulsed Laser Ablation Process," Applied Surface Science, Vol. 86, p. 213-18, Feb. 1995.

5. Ihlemann, J., Wolff-Rottke, B., "Excimer Laser Ablation Patterning of Dielectric Layers," Applied Surface Science, Vol. 86, p. 228-33, Feb. 1995.

6. Jackson, S.R., Metheringham, W.Z., Dyer, P.E., "Excimer Laser Ablation of Nd:YAG and Nd:Glass," Applied Surface Science, Vol. 86, p. 223-7, Feb. 1995.

7. Costela, A., Garcia-Moreno, I., Florido, F., Figuera, J.M., Sastre, R., Hooker, S.M., Cashmore, J.S., Webb, C.E., "Laser Ablation of Polymeric Mateials at 157 nm," Journal of Applied Physics, Vol. 77, No. 6, p. 2343-50, March 1995.

8. Kautek, W., Krueger, J., "Femtosecond-Pulse Laser Microtructuring of Semiconducting Materials," Materials Science Forum, Vol. 173-174, p. 17-22, 1995.

9. Tonshoff, H.K., Mommsen, J., "Process of Generating 3-D Microstructures Using Excimer Lasers," Laser und Optoelektronik, Vol. 24, No. 1, p. 64-7, Feb. 1992.

10. Sugioka, K., Toyoda, K., "Modification and Microfabrication of Solid Surfaces Using Excimer Lasers," Riken Review, No. 1, p. 31-2, April 1993.

11. Nutt, A.C.G., "One-Step Stripe Waveguide Fabrication Process for Polymer Integrated Optics," Proceedings of the SPIE – The International Journal for Optical Engineering, Vol 1794, p. 421-33, 1993.

12. Matz, R., Heydel, R., Gopel, W., "In Situ Fabrication of InP-Based Optical Waveguides by Excimer Laser Projection," Applied Physics Letters, Vol. 63, No. 8, p. 1137-9, Aug. 1993.

13. Christensen, C.P., Duignan, M.T., Rodriquez, L.R., "Micromachining with waveguide excimer lasers," SPIE, Vol. 1835, p. 128-32, 1993.

14. Kantsyrev, V.L., Komardin, O.V., Solinov, V.F., Yakovlev, M., "New methods of cleaving glass by ultraviolet laser radiation," Pis',a v Zhurnal Teknicheskoi Fizika, Vol. 20, No. 11-12, p. 89-92, 1994.

15. Guzzo, E.E., Preston, J.S., "Laser ablation as a processing technique for metallic and polymer layered structures," IEEE Transactions on Semiconductor Manufacturing, Vol. 7, No. 1, p. 73-8, 1994.

16. Tejedor, P., Briones, F., "Ultraviolet laser-projection patterning of polymeric materials for electrochemical gas sensors," Applied Physics Letters, Vol. 64, No. 7, p. 936-8, 1994.

17. Mori, M., Kemmochi, Y., Yamaguchi, S., Sekine, K., "Applications of excimer lasers for maching of fiber reinformced plastics," SPIE, Vol. 2118, p. 204-8, 1994.

18. Dickmann, K., Dik, E., "Laser drilling of fine holes," Technica, Vol. 43, No. 10, p. 59-65, 1994.

19. Christensen, C.P., "Waveguide excimer laser fabrication of 3D microstructures," SPIE, Vol. 2045, p. 141-5, 1994.

20. Leppavuori, S., Levoska, J., Hill, A.E., Tomlinson, R.D., "Laser ablation deposition as a preparation method for sensor materials," Sensors and Actuators A (Physical), Vol. A41, No. 1-3, p. 145-9, 1994.

21. Osvay, K., Bor, Z., Racz, B., Heitz, J., "Direct writing and in-situ material processing by a laser-micromachining projection microscope," Applied Physics A (Solids and Surfaces), Vol. A58, No. 3, p. 211-14, 1994.

9.8. WIRE STRIPPING AND FIBER TAPS

22. Harvey, E.C., Rumsby, P.T., Gower, M.C., Mihailov, S., Thomas, D., "Excimer lasers for micromachining," IEE Colloquium on 'Microengineering and Optics' (Digest No. 1994/043), p. 62, 1/1-4, 1994.

23. Leppavuori, S., Levoska, J., Hill, A.E., Tomlinson, R.D., Frantti, J., Kusmartseva, O., Moilanen, H., Pikington, R.D., "Laser ablation deposition as a preparation method for sensor materials," Sensors and Actuators, A: Physical Vol. 41, N. 1-3, Pt 3, p. 145-149, 1994.

24. Christensen, C.P., "Capabilities of low power excimer lasers in micromachining," Conference proceedings - Lasers and Electro-Optics Society Annual Meeting 1993, IEEE, p. 762-763.

25. Wolff-Rottke, B., Schmidt, H., Scholl, A., Ihlemann, J., "Micro machining with excimer lasers: photoablation and plasma sputtering," SPIE, Vol. 1810, p. 650-3, 1994.

26. Tonshoff, H.K., Hesse, D., Gedrate, O, "Gas-assisted microstructuring of ceramic materials irradiated with excimer lasers," SPIE, Vol. 1810, p. 572-6, 1994.

27. Goller, M., Lutz, N., Geiger, M., "Micromachining of ceramics with excimer laser radiation," Journal of the European Ceramic Society, Vol. 12, No. 4, p. 315-21, 1994.

28. Christensen, C.P., Duignan, M.T., Rodriquez, L.R., "Micromachining with waveguide excimer lasers," SPIE, Vol. 1835, p. 128-32, 1993.

29. Tabat, M., O'Keeffe, T.R., Ho, W., "Profile characteristics of excimer laser micromachined features," SPIE, Vol. 1835, p. 144-57, 1993.

30. Gower, M.C., "Industrial micromachining applications of excimer lasers," IEEE Colloquium on 'Laser Applications' (Digest No. 084), p. 52, 7/1-3, 1993.

31. Damiani, D., Catherinot, A., "Excimer laser micromachining," Annales de Physique, Vol. 17 colloq, No. 1, p. 57-8, 1993.

32. Ihlemann, J., Schmidt, H., Wolff-Rottke, B., "Excimer laser micromachining," Advanced Materials for Optics and Electronics, Vol. 2, No. 1-2, p. 87-92, 1993.

Chapter 10
UV Lasers in Medicine

10.1 Introduction

UV lasers are emerging as useful tools in the medical field. UV lasers have been applied to a wide range of medical applications, and several have matured to become clinically significant. In this chapter, we will present examples of how ultraviolet light is being successfully applied to solve medical problems or to provide an improved method of treating human disorders.

10.2 Applications

The uses of UV laser technology in the medical field are increasing as new optical delivery systems become available. The primary applications include angioplasty (plaque removal from interior walls of arteries), 193-nm micromachining of bone for insertion of microminaturized electronic hearing devices, cornea tissue ablation at 193 nm to correct vision and remove surface lesions and tumors, dental surgery by ablation of tooth surfaces, fluorescence research in microbiology to identify absorption mechanisms in cell chemistry, and skin surgery to remove unwanted surface tissue structures.

In addition to UV ablation reactions conducted on the surface of the human body, fiber-optical delivery of UV radiation inside the body provides a precise cutting tool. For example, tumor and gall stone removal by ablation processes eliminates the need for invasive surgery.

DNA and related research in microbiology is possible by targeted UV ablation and structuring of individual cells and portions of cells.

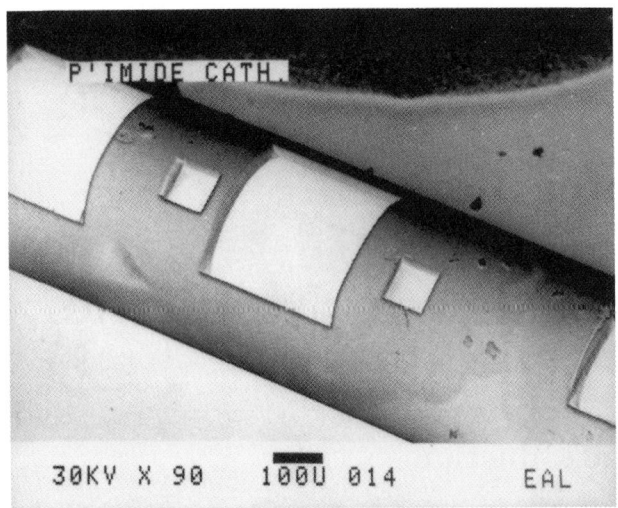

Figure 10.1: T-Joint Sites Ablated in a Polymeric Catheter

Intravenous dispensing of medications under computer control is facilitated with catheters that are fabricated by UV laser ablation. One example of a 193-nm micromachined polymeric catheter is shown in Figure 10.1 (20).

The high resolution imaging and ablation features of UV laser technology are also used in neurosurgery for clean, cross-sectional ablation of nerve fibers to permit the rejoining and regeneration of a damaged nerve.

10.3 UV Laser Advantages in Medicine

There are several functional advantages of UV lasers in medicine that cannot be provided by longer wavelength sources or other energy sources. These include:

Athermal Reactions: The absorption of UV photons on the surface (500-1500 Å) of skin or bone tissue results in the ablative removal of both material and heat. The material below the ablated layer (bulk) is largely unaffected thermally. This property is often critical in microsurgery. The photos in Figures 10.2 and 10.3 illustrate the difference between samples of collagen which have been laser drilled with carbon dioxide, Nd:YAG, and UV lasers (19).

10.3. UV LASER ADVANTAGES IN MEDICINE

CO_2-laser drill	Nd:YAG-laser drill
laser power : $P_L = 500$ W pulse frequency : $f_p = 1$ Hz	pulse energy : $E_p = 1{,}2$ J pulse frequency : $f_p = 1$ Hz
material : Collagen (4 mm thick)	

Figure 10.2: CO_2 and Nd:YAG Laser Drilling of Collagen

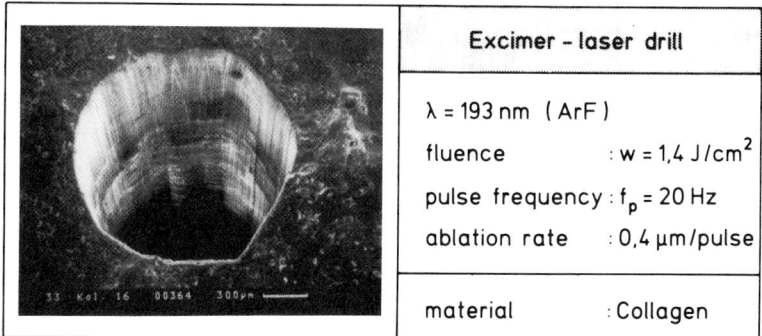

	Excimer-laser drill
	$\lambda = 193$ nm (ArF) fluence : $w = 1{,}4$ J/cm^2 pulse frequency : $f_p = 20$ Hz ablation rate : $0{,}4$ µm/pulse
	material : Collagen

Figure 10.3: Excimer Laser Drilling of Collagen

Projection Imaging: A UV laser may be optically focused and projected over some distance to the desired site. UV optical surgery by projection imaging keeps the instrument from touching the patient, reducing the danger of infection from bacteria.

Visual Targeting: Visible wavelengths can be threaded coaxially along with UV beams by using mirror-based optical delivery systems. Prior to firing the laser, a surgeon can preilluminate the area to be cut or ablated, enlarge or reduce the area, shape, or size, and then initiate the UV laser pulses for the actual ablation. Previewing in medical surgery, where you can see what you are operating on in this manner, is a significant advantage, especially as the light allows increased vision at high resolution and magnified on the site area at high magnification.

Precise Depth Removal: UV laser ablation of tissue takes place one precise layer at a time, in layer thicknesses as small as 1000 Å or less. In medicine, removal of material down to a specific boundary or stopping point (for example, adjacent to another organ) can be important. Individual laser pulses can be regulated by adjusting pulse energy through the laser or through optical attenuators. Also, the number of pulses can be selected to achieve a precise amount of tissue removal. Precise depth removal can also be affected by the laser wavelength. Each wavelength absorbs differently, and cutting depth is therefore different. Available wavelengths for medical laser delivery systems range from 351 to 193 nm.

10.4 UV Laser Angioplasty

Removal of arterial buildup is a major application for lasers. Argon lasers are used for thermal removal of atrial plaque, and excimer lasers are used for nonthermal removal. Angioplasty is a medical procedure designed to reduce ischemic artery disease. The consequences of this disease, left unchecked, are heart attack, stroke, amputation, and death.

The conventional medical procedures to treat this disease are either coronary bypass surgery or balloon angioplasty, where a catheter with a balloon on its tip is inserted into the artery. The balloon is inflated at the desired location, and the pressure of the balloon causes redistribution of the buildup of material lining and partially clogging the artery.

Argon laser angioplasty, a thermal-removal technique, was used initially as a secondary or complementary procedure along with balloon angioplasty. Laser angioplasty, with argon and UV lasers, is now a stand-alone procedure.

10.4. UV LASER ANGIOPLASTY

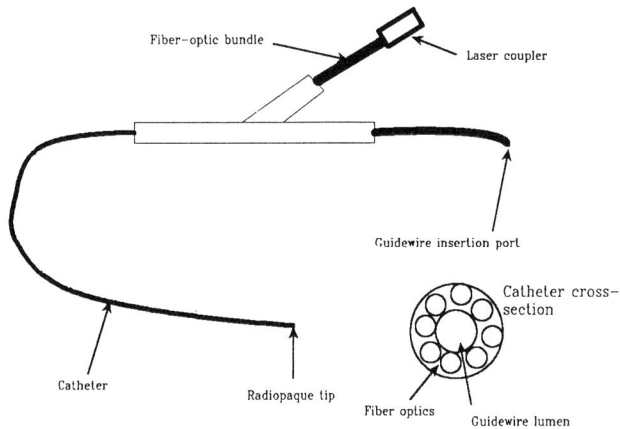

Figure 10.4: Multifiber Excimer Laser-based Angioplasty System

The improvements of the delivery optics and technique for laser angioplasty have improved considerably with time.

UV laser angioplasty has developed slowly because of the lack of adequate optical delivery technology. UV energy must be transmitted through a fiber optical delivery system, and highly transmissive silica fibers are needed to make the procedure practical. Also, a means to perform this procedure without damaging the healthy walls of the artery is needed. Figure 10.4 shows an example of a laser angioplasty system which is multifiber excimer laser-based (1).

In early work in this area, xenon chloride excimer lasers with fiber optical delivery systems were used to recanalize coronary arteries in open heart surgical procedures. This was followed by percutaneous fiber optical surgery for the same operation. A xenon chloride excimer laser at 308 nm and a xenon fluoride excimer laser at 351 nm are relatively long deep-UV wavelengths, which are not as efficient at ablating organic material as are the shorter excimer laser wavelengths, such as krypton fluoride at 248 nm or argon fluoride excimer lasers at 193 nm. However, early fiber optical delivery systems did not provide the damage-threshold resistance required for the shorter deep-UV wavelengths, where the desired high photon energy can be used (1).

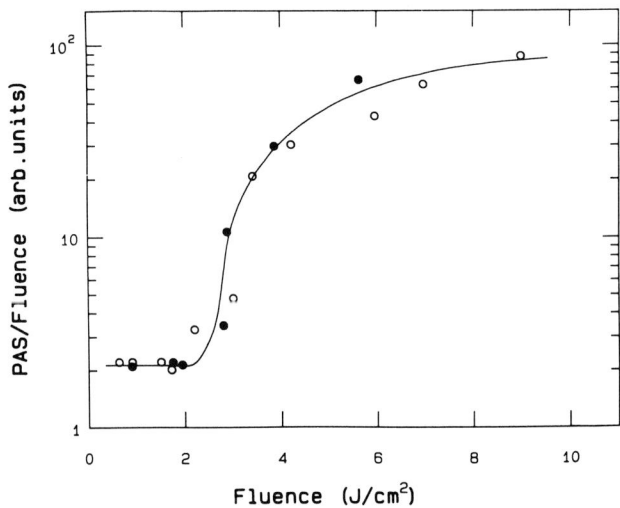

Figure 10.5: Normalized Photoacoustic Signals for Human Aorta as a Function of Incident Fluence

Fiber delivery in the deep-UV range requires selection of special fiber characteristics, an efficient means of coupling the laser energy from the source through a fiber to the patient, and getting sufficiently long laser pulses for optimum results. Fiber damage and fiber transmission are key issues, and the optical delivery system must provide very uniform laser energy distribution and low loss to result in a useful medical tool. The use of very long excimer laser pulses shows that much higher damage thresholds can be obtained without any significant disadvantages.

Figure 10.5 shows the photoacoustic spectroscopy signal of a human aorta as a function of XeCl laser fluence for optical pulse durations of 7 and 300 nsec. The photoacoustic signal reaches the target first, followed immediately by the onset of photoablation where the sharp rise in photo acoustical signal/fluence occurs (5).

The data show that the ablation threshold for the longer 300-psec pulse length, which is desired for fiber optical transmission, is only about 25% greater than the threshold for the short 7-nsec pulse length. Ablation of dielectric films such as Kapton and Mylar have shown similar results (8).

10.4. UV LASER ANGIOPLASTY

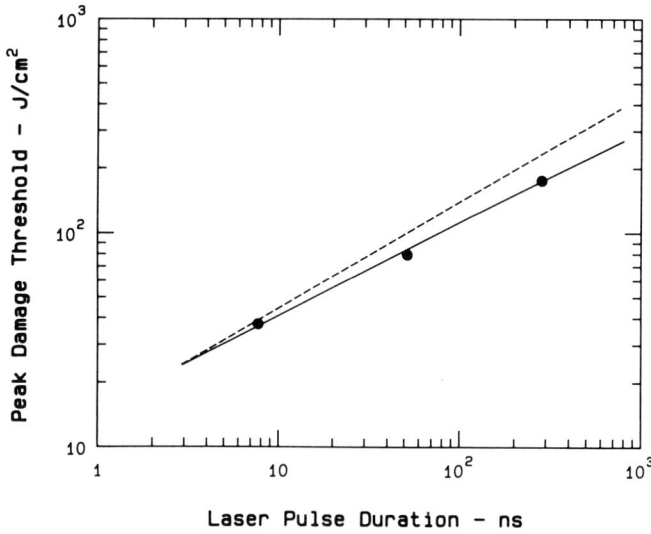

Figure 10.6: Catastrophic Front Surface Damage Thresholds as a Function of XeCl Laser Pulse Duration

Once it was understood that there is only a small loss in efficiency for using long excimer laser pulse lengths, the actual damage data and nonlinear transmission behavior was studied. Several researchers explored this area, and the results showed that the damage threshold increases with optical pulse duration. Figure 10.6 shows the peak catastrophic fiber surface damage as a function of XeCl laser pulse duration. This study used Disagreed ST-U fibers with a 400-μm core diameter.

Optimizing a fiber optical beam delivery system for an excimer laser also requires looking at parameters such as intensity dependence on color center formation, surface quality of the fused silica, uniformity of the laser beam, wavelength dependence, and other nonlinear effects. For example, nonlinear effects include stimulated Brillouin scattering, stimulated Raman scattering, self-focusing, color center formation, and multiphoton absorption. In a fiber optical delivery system with average (<10 m) fiber lengths, the most significant nonlinear effect is multiphoton absorption (6). The two-photon absorption coefficients for fused silica at various excimer laser wavelengths play a key role in damage behavior.

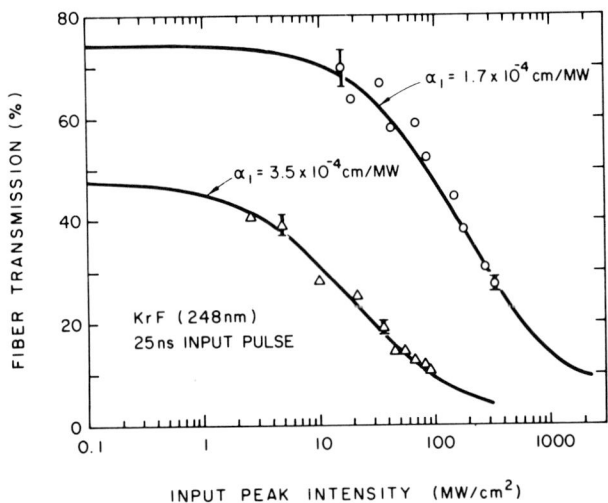

Figure 10.7: Fiber Transmission vs Input Peak Intensity

The two-photon absorption coefficients (TPA) for xenon fluoride and xenon chloride are 4.6 X 10^{-7}; the values for krypton fluoride and argon fluoride are 2.1 and 2.0 respectively (5). Longer pulse lengths will reduce the TPA, and higher purity materials will also improve the results.

Computer simulation of the transmission of excimer laser energy through fused silica fibers shows a significant reduction in fiber transmission as input peak intensity increases. Figure 10.7 illustrates this relationship. This study used a Disagreed ST-U fiber with 350- and 62-cm lengths.

The two-photon absorption (TPA) coefficients are so large at the very short excimer laser wavelengths (248 and 193 nm) that significant attenuation occurs. Current fiber optical delivery systems use reduced laser intensity, closer to the ablation threshold of the tissue material, and typical pulse durations of 15-25 nsec. These parameters permit optical transmission in the range of 60-75% with fibers up to 3-4 feet in length.

Improvements in fused silica fiber technology and optimization of overall system parameters (uniformity, fiber quality, optimum pulse length) have permitted the use of 193-nm optical delivery systems. The high photon energy at 193 nm and intense absorption result in a clean ablation source, and material is reacted into gaseous and molecular by-products with few solid fragments.

10.5. CORNEAL SCULPTING

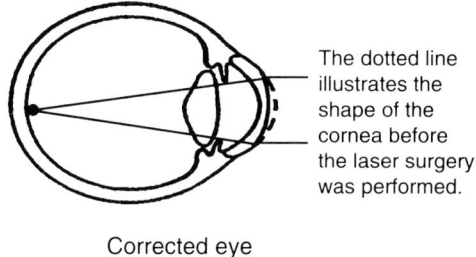

Corrected eye

Figure 10.8: Human Eye Schematic before (dotted line) and after PRK

10.5 Corneal Sculpting

The use of an excimer laser to perform a photorefractive keratectomy (PRK) to correct myopia is well documented and is used clinically in many countries in the world. Photorefractive keratectomy is a surgical procedure developed to correct myopia by structuring corneal tissue, through selective removal. An excimer laser is used to photoablate the tissue over a predetermined area, with the area and depth determined by the amount of vision correction needed. Typically, a 193-nm laser is used in conjunction with a fiber optical delivery system. The laser energy is shaped and delivered to the cornea, where intense absorption results in ablation and vaporization of successive thin layers of tissue until the prescribed depth is reached. The entire procedure takes only minutes and can be performed with little to no anesthesia. The result is a flattening of the cornea which reduces the myopia. Figure 10.8 shows a schematic of the human eye before and after this procedure is performed (2).

In a normal eye, vision is the result of light rays being focused by the cornea and the lens onto the retina, a layer of light-sensitive tissue lining the back of the eye. The retina converts the light into signals for the brain to process. Images that are focused precisely on the retina permit good vision without glasses or contact lenses in a healthy eye.

There are a number of refractive errors in human vision, including myopia, hyperopia, and astigmatism, each of which have the problem of focusing the light either in front of or "behind" the retina. In myopia, the light focuses just in front of the retina and is often caused by the eyeball being elongated, as shown in Figure 10.9, where it is compared to a normal eye.

Myopia is also caused by the cornea being highly curved or unusually steep. A highly curved cornea also causes images to focus in front of the

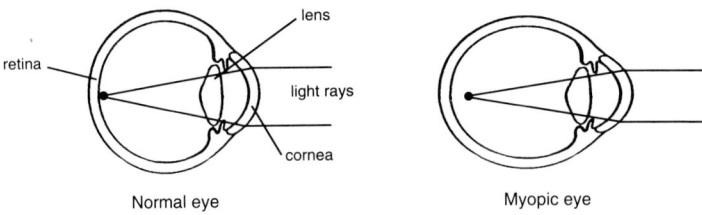

Figure 10.9: Schematic of a Normal and a Myopic Eye

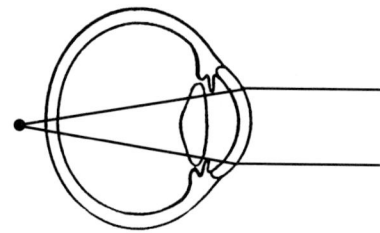

Figure 10.10: Schematic of a Hyperopic Eye

retina, even with a normally shaped eyeball. Hyperopia, or farsightedness, occurs when light rays are focused at a point behind the retina. People with hyperopia have difficulty seeing objects close up. A hyperopic eye is shown in Figure 10.10 (2).

Astigmatism usually is the result of an irregularly shaped cornea. A normal cornea is shaped like a basketball, whereas an astigmatic cornea is shaped more like a football, causing light rays to focus at different points rather than on a single point. Astigmatism makes vision blurred and can distort objects.

In excimer laser eye surgery, UV light is used to remove precise amounts of corneal tissue to reshape the "optical system" of the human eye. Corneal tissue removal is accomplished without depositing any significant heat on the eye itself, and the ablation reaction carries away both the unwanted corneal tissue and the excess heat generated in the reaction.

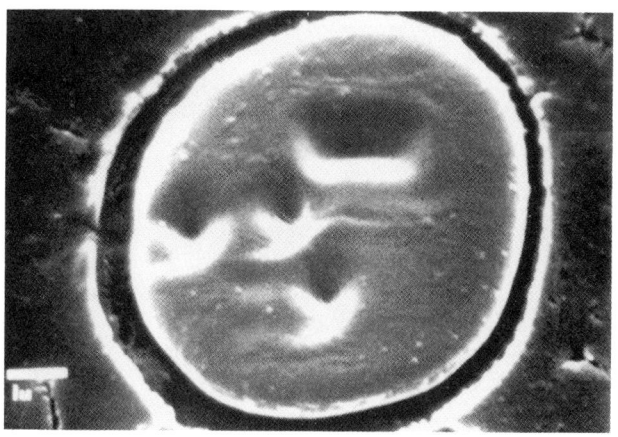

Figure 10.11: Human Blood Cell with UV Laser Cuts

10.6 Microbiology Research

In microbiology research, the UV laser can be used to create openings in individual cells. UV laser beams can be focused to extremely high resolution spot sizes, typically less than 1 μm. These highly focused spots are imaged directly onto targeted sites, either the surface of a blood cell or a section of a large molecule, such as DNA.

In Figure 10.11, an example of blood cell ablation is shown. An excimer laser at 193 nm was focused through a Schwarzchild mirror lens, and several pulses were fired directly onto the cell surface. The result is ablation of the membrane of the cell wall, creating an opening in the cell for the introduction of various experimental materials or chemistry.

The excimer laser may also be used to slice cross-sectionally through a blood cell or other biological structure, such as an egg or zygote. In cattle breeding, for example, it would be useful to extract a fertilized egg from a prize animal, section the egg into several pieces, and reinsert these pieces into several host animals to produce identical offspring. Figure 10.12 shows a human blood cell cut cross-sectionally with an excimer laser at 193 nm. Approximately nine pulses were used to cut this cell (20).

The capability to gain access to individual molecules, cells, and other

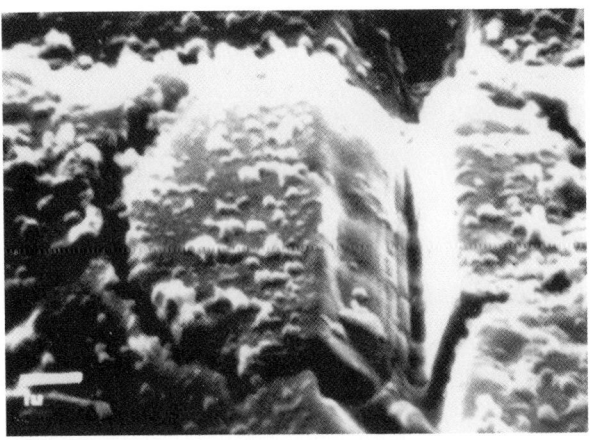

Figure 10.12: Human Blood Cell Sectioned with a 193-nm Excimer Laser

biological structures without changing their internal chemistry and without damaging physically or thermally the workings or structure of the organism is extremely valuable in medical research. Current methods rely on physical manipulation and puncturing which often distorts and damages the structure chemically. Etching through cells can introduce undesirable side effects and limits the value of the research experiment. Deep-UV radiation, focused into a high resolution spot, permits photoablation and removal of precise amounts of organic material, without causing this damage.

An optical delivery system that permits the imaging of various deep-UV wavelengths down to submicron spot sizes is shown in Figure 10.13. This optical system (U.S. patent no. 5,206,515) has been used to create submicron structures and images for microlithography, having resolution requirements down to 0.25 μm (20).

This system allows visible light to be first targeted onto the specimen, permitting the operator to preview and even shape the area where the UV laser will be directed for ablation. On the same optical path, the UV energy is imaged through a 10x reduction lens to the target. A wide range of pulse energies, laser wavelengths, and doses can be delivered while the researcher watches the reaction.

As new optical delivery systems become available for this application, this

10.7. LASER MICROSURGERY

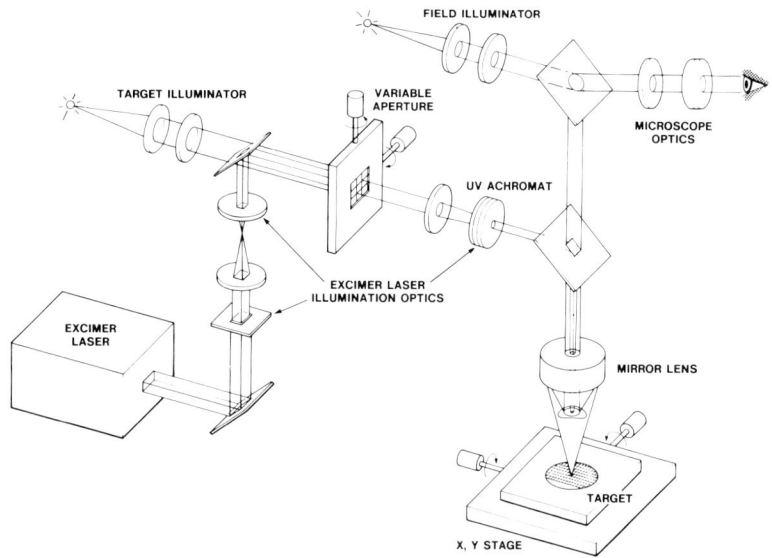

Figure 10.13: Deep-UV Optical Delivery System for Submicron Imaging

type of research can expand. For example, Figure 10.14 shows a computer-aided model of a treatment algorithm for excimer laser human eye correction of both astigmatism and hyperopia in the same operation (16).

10.6.1 Beam-Scanning Techniques for PRK

Large area ablation has been the dominant technnique for corneal sculpting or PRK. One of the issues with large are ablation is haze or scarring that remains in some patients. The use of a small scanning beam is believed to leave a smoother surface, provided good eye movement tracking can be provided. The computer-aided model of a treatment alogrithm for simultaneous correction of both astigmatism and hyperopia is expected to replace non-scanning methods.

10.7 Laser Microsurgery

Microsurgery and microstructuring with excimer lasers have numerous applications in medical research and in clinical facilities. Medical applications

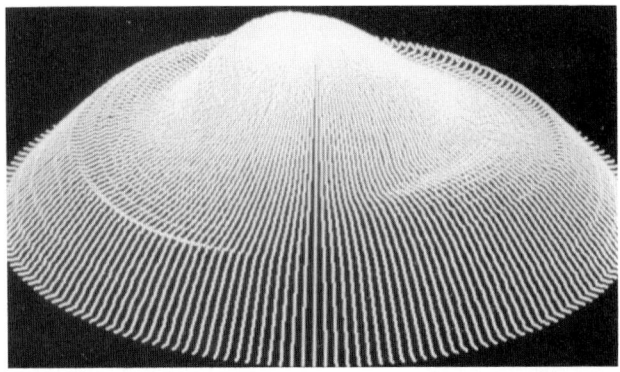

Figure 10.14: Computer Model of Excimer Beam Scan of the Eye to Correct Astigmatism and Hyperopia (Courtesy of Aesculap/Meditec)

for laser microsurgery typically pose the problem of imaging into highly absorbing bio-fluids or tissues which are predominantly composed of water, and positioning the laser beam delivery tool with control in three to four axes of motion.

Figure 10.15 shows an example of how these problems were solved by using a finely tapered capillary, fed with a stream of positive pressure gas, and positioned on the end of an articulated arm high efficiency, high damage threshold material which in turn is connected to an excimer laser. The left side of the photo shows an ordinary pipette supporting a mouse oocyte center of photo (19).

The laser drilling apparatus described was used for in vitro pellucida or outer covering of a mouse oocyte for in vitro fertilization. The UV laser drilling of holes in the oocyte permits male sperm to more easily penetrate the egg. The glass-tapered tip that directs the laser energy allows micron-level precision and holes are drilled small enough to keep more than one sperm from entering (19).

Figure 10.16 shows a) a 2000x magnification SEM of holes bored in the mouse vocyte; and b) a close-up of the same sample at 5000x magnification. The 193-nm laser fluence was 5-8 mJ/cm^2. The energy output at the tip of the glass pipette was only 1 mJ.

10.7. LASER MICROSURGERY

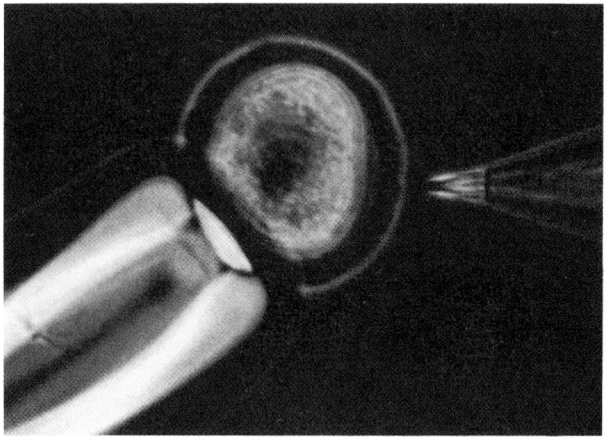

Figure 10.15: Mouse Oocyte (Center), Held by Pipette (Left) Ready for ArF Excimer Laser Drilling by the Air-Filled Glass Micropipette (Right) (Courtesy of Lambda Physik)

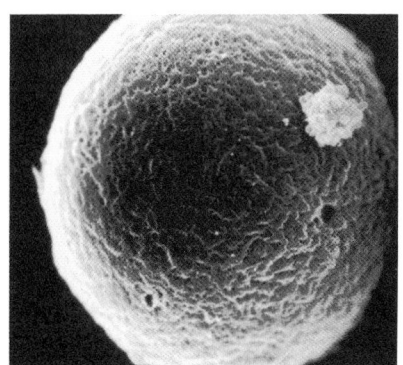 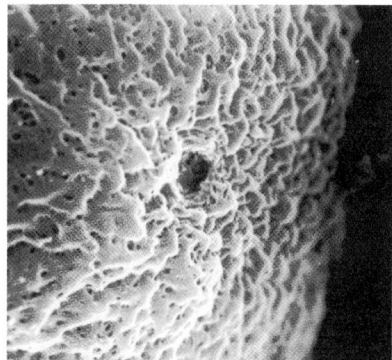

Figure 10.16: Mouse Oocyte with Laser Ablated Opening; SEM at a) 2000 mag, b) 5000 mag

Figure 10.17: Medical Bilumen Catheter with 500-nm Hole Made with 248-nm Radiation (Courtesy of Excitech, Ltd.)

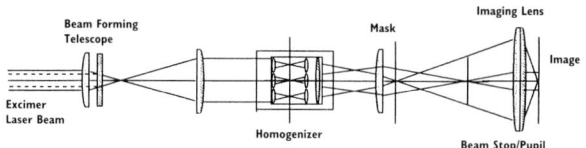

Figure 10.18: Mask Illumination and Imaging Optics (Courtesy of MicroLas)

10.8 Catheter Machining Optics

Medical catheters are widely used for fluid delivery and other patient support systems. Micromachining of small (200-500 μm) holes in the ends of fibers is a requirement, and Figure 10.17 shows a bilumen catheter machined with a 248-nm excimer laser. The catheter is used to measure the oxygen content in the blood of premature babies (20).

A typical example of optics used to machine small structures is illustrated in Figure 10.18. The laser beam is first shaped with a beam-forming telescope, then processed through a homogenizer and a mask creating a uniformly illuminated aerial image. This image is formed onto the substrate with a demagnifying lens. The final lens can be either a refractive-type

10.8. CATHETER MACHINING OPTICS					333

all-quartz microscope objective or a simple two-mirror reflecting objective (20).

References

1. Dullien, R. C., "Medical Laser Applications," OE Reports, January 1991.

2. "Correction of Nearsightedness with the excimer laser," Massachusetts Eye and Ear Infirmary, March, 1995.

3. Taylor, R.S., Leopold, K.E., Brimachmbe, R.K., Keon, W.J., Higginson, L.A.J., "Fiber Optic Delivery System for the Excimer Laser Recanalization of Human Coronary Arteries in Open Heart Surgery," SPIE Vol. 906, pp. 225-230, 1988.

4. Grundfest, W., "Development of Percutaneous Excimer Laser Angioplasty Systems for Peripheral and Coronary Arteries," paper UV5.1, LEOS Annual Meeting, November 2-4, 1988, Santa Clara, California.

5. Taylor, R. S., Leopold, K.E., Brimaconbe, R. K., "Long Optical Pulse Excimer Lasers for Fiber Optic Delivery," SPIE Vol. 1041, pp. 198-203, 1989.

6. Taylor, R.S., Leopold, K.E., Brimacombe, R.K., Mihailov, S., "Dependence of the Damage and Transmission Properties of Fused Silica Fibers on the Excimer Laser Wavelength," Appl. Opt. 27, pp. 3124-3134, 1988.

7. Taylor, R.S., Leopold, K.E., Mihailov, S., Brimacombe, R.K., "Damage Measurements of Fused Silica Fibers Using Long Optical Pulse XeCl Lasers," Opt. Commun. 63, pp. 26-31, 1987.

8. Taylor, R.S., Singleton, D.L., Paraskevopoulos, G., "Effect of Optical Pulse Duration on the XeCl Laser Ablation of Polymers and Biological Tissue," Appl. Phys. Lett. 50, pp. 1779-1781, 1987.

9. Nevis, E.A., "Alteration of the Transmission Characteristics of Fused-Silica Optical Bibers by Pulsed Ultraviolet Radiation," SPIE 540, pp. 421-426, 1985.

10. Arai, K., Imai, H., Hosono, H., Abe, Y., Imagawa, H., "Two-Photon Processes in Defect Formation by Excimer Lasers in Synthetic Silica Glass," Appl. Phys. Lett. 53, pp. 1891-1893, 1988.

11. Escher, G., "KrF Laser Induced Color Centers in Commercial Fused Silicas," Proc. of SPIE Symposium on Excimer Interaction with Materials, Sept. 6-9, Boston, MA, 1988.

12. Sliney, D.H., "The Need for Dosimetry in Medical Treatment," SPIE Institute Series Vol. IS-5, pp. 5-8, 1989.

13. Brunner, T.M., "Laser Industry Looks at Laser Medicine and Surgery," SPIE Vol. 357, Lasers in Medicine and Surgery, 1982, pp. 7-10.

14. Leckrone, M., "Advances in Laser Angioplasty," SPIE Vol. 953, The Marketplace for Medical Lasers (1988), pp. 33-40.

15. Sanborn, T. et al, "Coronary Laser Angioplasty: Reduced Complications and Indium-111 Platelet Accumulation Compared with Thermal Laser Angioplasty," Journal of the American College of Cardiology, Vol. 16, No. 2, Aug. 1990, pp. 502-506.

16. "An Eye for Tommorrows Applications Spells Success in Medical Lasers," Photonics Spectra, October, 1994.

17. Lambda Physik "Highlights" No. 18, August, 1989.

18. Image Micro Systems, Application Photos, October, 1988.

19. Lambda Physik "Highlights" No. 38 and No. 39, February, 1993.

20. Lambda Physik "Highlights" No. 8, November, 1994.

Glossary of Terms

Angioplasty Referring to the repair of an artery, in this case by using UV laser energy to remove plaque from sidewalls.

Catheter (Polymeric) A flexible tubular structure used to deliver fluids inside a human, generally for either intravenous feeding or medicating. A polymeric catheter is a catheter made from a liquid polymer material.

10.8. CATHETER MACHINING OPTICS

Damage Threshold A specific input energy level, generally radiant laser energy, which produces sufficient absorption in a lens or fiber optical material to cause solarization, color centers, or preablation sites, all of which reduce UV transmission and are considered as damage.

Fiber Laser A laser in which the lasing medium is an optical fiber doped with low levels of rare earth halides to permit the amplification of light. The output of a fiber laser can be tunable over a wide range or can be broadband. Fiber laser can be pumped by diode lasers.

Fiber Optics The area of optics technology that deals with the transmission of radiant power through UV (or other wavelengths) transmissive fibers made of quartz, UV transmitting plastics, or glass.

Hyperopic Refers to farsightedness when the eye focuses light rays behind the retina instead of directly on the retina.

Myopia Refers to nearsightedness when the eye focuses light rays in front of the retina. Myopia occurs when the cornea or eyeball is highly curved.

Photoacoustic (Effect) The generation of a sound-based signal caused by modulated light irradiation.

PRK For photorefractive keratectomy. A surgical procedure on the cornea of the eye wherein selective portions are removed. For example, in a radial keratectomy, pie-shaped slices are removed in a radial pattern to correct the myopia.

Reading Bibliography

1. Lin, J.T., "Critical Review on Refractive Surgical Lasers," Optical Engineering, Vol. 34, No. 3, p. 668-75, March 1995.

2. Pettit, G.H., Ediger, M.N., Weiblinger, R.P., "Excimer Laser Ablation of the Cornea," Optical Engineering, Vol. 34, No. 3, p. 661-7, March 1995.

3. Malo, B., Albert, J., Hill, K.O., Bilodeau, F., Johnson, D.C., Theriault, S., "Enhanced Photosensitivity in Lightly Doped Standard Telecommunication Fibre Exposed to High Fluence ArF Excimer Laser Light," Electronics Letters, Vol. 31, No. 11, p 879-80, May 1995.

4. "International Conference on Advanced Laser Dentistry" Proceedings of the SPIE – The International Journal for Optical Engineering, Vol. 1984, 1995.

5. "Laser Technology IV: Applications in Medicine," Proceedings of the SPIE – The International Journal for Optical Engineering, Vol. 2203, 1995.

6. Hjortdal, J.O., Erdmann, L., Bek, T., "Fourier Analysis of Video-Keratographic Data. A Tool for Separation of Sperical, Regular Astigmatic and Irregular Astigmatic Corneal Power Components," Ophthalmic and Physiological Optics, Vol. 15, No. 3, p. 171-85, May 1995.

7. Nakayama, T., Kubo, U., "Effect of UV Laser Irradiation on Tissue," Review of Laser Engineering, Vol. 20, No. 11, p. 845-53, Nov. 1992.

8. Raynal, E., Berthier, J.P., Barbe, R., Kime, S., "Action of Excimer Laser Radiation at 308 nm on Myocardial Cells and on Mitochondria," Lasers in the Life Sciences, Vol. 5, No. 3, p. 165-74, 1993.

9. Gower, M.C., Rumsby, P.T., Thomas, D.T., "Novel Applications of Excimer Lasers for Fabricating Biomedical and Sensor Products," Proceedings of the SPIE – The International Journal for Optical Engineering, Vol. 1835, p. 133-42, 1993.

10. Pettit, G.H., Saidi, I.S., Tittel, F.K., Sauerbrey, R., Cartwright, J., Jr., Farrel, R., Benedict, C.R., "Thrombolysis by Excimer Laser Photoablation," Lasers in the Life Sciences, Vol. 5, No. 3, p. 185-97, 1993.

11. Pettit, G.H., Ediger, M.N., Weiblinger, R.P., "Dynamic Optical Properties of Collagen-Based Tissue During ArF Excimer Laser Ablation," Applied Optics, Vol. 32, No. 4, p. 488-93, Feb. 1993.

12. Feller, K.H., Clausner, G., Konig, R., "Experimental Studies on UV Laser Ablation of Tissues," Experimentelle Technik de Physik, Vol. 39, No. 4-5, p. 319-22, 1991.

13. Neev, J., Raney, D.V., Whalen, W.E., Fujishige, J.T., Ho, P.D., McGrann, J.V., Berns, M.W., "Dentin Ablation with Two Excimer Lasers: A Comparative Study of Physical Characteristics," Lasers in the Life Sciences, Vol. 5, No. 1-2, p. 129-53, 1992.

14. Rimoldi, D., Flessate, D.M., Samid, D., "Changes in gene expression by 193- and 248- nm excimer laser radiation in cultured human fibroblasts," Radiation Research, Vol. 131, No. 3, P. 325-31, 1992.

15. Hillrichs, G., Dressel, M., Hack, H., Kunstmann, R., Neu, W., "Transmission of XeCl excimer laser pulses through optical fibers: dependence on fiber and laser parameters," Applied Physics B (Photophysics and Laser chemistry), Vol. B54, No. 3, P. 208-15, 1992.

16. van Leeuwen, T.G., Jansen, E.D., Motamedi, M., Welch, A., Borst, C., "Excimer laser ablation of soft tissue: A study of the content of rapidly expanding and collapsing bubbles," IEEE Journal of Quantum Electronics, Vol. 30, No. 5, P. 1339-1344, 1994.

17. Hahn, D.W., Ediger, M.N., Pettit, G.H., "Plume particle dynamics during ArF excimer corneal ablation," Conference Proceedings - Lasers and Electro-Optics Society Annual Meeting, Vol. 8, Publ. IEEE, 1994.

18. Moss, J.P., Patel, B.C., Pearson, G.J., Arthur, G., Lawes, R.A., "Krypton fluoride excimer laser ablation of tooth tissues: precision tissue machining," Biomaterials, Vol. 15, No. 12, P. 1013-18, 1995.

19. Fujisaka, S., Ito, T., Sato, K., "High-speed imaging study of excimer laser ablation of bone," Review of Laser Engineering, Vol. 22, No. 7, P. 552-8, 1995.

20. Kautek, W., Mitterer, S., Krueger, J., Husinsky, W., Grabner, G., "Femtosecond-pulse ablation of human corneas," Applied Phyusics A: Solids and Surfaces, Vol. 58, No. 5, P. 513-518, 1994.

21. Pettit, G.H., Ediger, M.N., Weiblinger, R.P., "Dynamic optical properties of collagen-based tissue during ArF excimer laser ablation," Applied Optics, Vol. 32, No. 4, P. 488-493, 1993.

22. Bor, Z., Hopp, B., Racz, B., Szabo, G., Marion, Z., Vincze, F., Ratkay, I., Mohay, J., Suveges, I., Fust, A., "Study of the ArF excimer laser ablation of the cornea," SPIE, Vol. 1983, Pt 2, P. 902-904, 1993.

23. Gower, M.C., Rumsby, P.T., Thomas, D.T., "Novel applications of excimer lasers for fabricating biomedical and sensor products," SPIE, Vol. 1835, P. 133-42, 1993.

24. Pokora, L., "Excimer lasers and their applications in industrial technology and in medicine," Opto-Electronics Review, No. 1, P. 13-20, 1993.

25. Avrillier, S., Tinet, E., Ettori, D., "Progress in the use of excimer lasers in medicine," Annales de Physique, Vol. 17, colloq, No. 1, P. 13-20, 1993.

Index

Ablation
- aluminum oxide — 296, 299
- applications for — 292-295, 308-310
- bond breaking — 37, 49
- catheter — 318, 332
- cells — 327-328
- ceramic — 296
- characteristics — 46-53
- cleaning — 199
- collagen — 319
- copper — 294
- delicate structures — 110
- eye — 325-326
- fiber — 310
- history — 34-36
- mechanisms — 36-41
- microelectromechanical systems (MEMS) — 295
- molybdenum — 297
- parameters — 44-46
- phenomenon — 33-34
- photoresist — 290
- plastic — 294
- plumes — 38-40
- reaction — 51, 290
- sapphire — 298, 300
- shockwave — 45-46
- silicon dioxide — 37
- spectral analysis — 95-98
- teflon — 53
- theory — 33
- threshold — 41-43, 145, 292, 295
- wavelength — 290
- wire — 308, 310

Absorption
- calculation — 41
- centers in optics — 126-127
- optical materials — 40-41, 124

Index

Absorption (continued
- polymer — 50
- of typical films — 42

Alexandrite laser — 24-26

Aluminum
- coatings — 141-143
- deposition — 142
- mirror conditioning — 144
- planarization — 230, 238-241
- spectral analysis — 95-98

Angioplasty — 320-322

Annealing
- electron beam — 213
- laser — 214-217
- parameters — 210-213
- technology — 209-213

Annular illumination — 255

Anti-reflection
- calculation — 130-134, 136-238
- coatings — 134-138, 230-231

Argon Fluoride — see Excimer laser

Athermal — 318, see also Ablation

Attenuator — 146-148

Bandwidth — 267-269

Barium fluoride — 124

Beam profiles — 76-78, 81, 232

Beam shaping — 232, 234-235, see Optical delivery systems

Beamsplitter — 148-149

Blood cells — 327-328

Bond breaking — 49

Calcium fluoride — 124, 126

Catheter — 332-333

Ceramic — 279-299

Cleaning
- compact discs — 187-191
- equipment — 180-181, 202-205

Index

Cleaning (continued)
    flat panel displays   184-187
    integrated circuits   171-181, 196-199
    mechanisms   199-202
    photoreactive   204-205
    printed circuits   191-196
    semiconductors   196, 205
    surfaces   173-177
    technology   173-177
    thin film heads   181-184
Coatings
    aluminum   141-143
    antireflection   143-138
    beamsplitter   148-151
    damage   151-156
    manufacturing   128-129
    multiple wavelength   150
    silicon dioxide   143-144
    uv optical   124-151, 134-138
Coherence   266-267
Collagen   319
Color centers   see Ultraviolet damage
Compact discs   187-191
Corneal sculpting   325-327

Damage   151-158, 323
Deep uv spectrum   2
Defects   197
Deposition   103
Dichroic beamsplitter   133
Dielectric coatings
    anti-reflection   134-138
    reflection   130-136, 138-145
Doping   217-219

Excimer laser
    beam profiles   76-77
    bond breaking   49
    cleaning   107-109
    damage   151-154, 323

Excimer laser (continued)
- deposition — 103
- discharge chamber — 73, 262
- energy per photon — 69
- electrical system — 262-263
- gain generator — 68
- gain profile — 70, 264, 270
- gases — 68, 74-75, 79-80
- history — 67-68
- imaging — 104, 106-107
- induced fluorescence — 100-102
- installation — 82
- maintenance — 82-85, 270-272
- mirrors — 148-151
- operation — 75-80, 83-85
- optics — 85
- preionizer — 72
- pulses — 77-79, 81
- reaction chamber — 69
- reactions — 69
- reliability — 80-81, 83-85
- resonator — 68, 73
- safety — 85-88
- subsystems — 71-75
- surface analysis — 104
- surface reactions — 49
- theory — 17, 68-71
- tunable — 101-102
- wavelengths — 17
- x-ray production — 99-100

Extreme ultraviolet (xuv) — 131-132, see Ultraviolet

Failure analysis — 308-309

Fiber
- optical delivery — 158-159, 320-325
- uv transmission — 322-324

Flat panel displays — 184-187

Fluence — 229, 292, 319, 322

Fluorine
- gas for excimer laser — 86

Index

Fluorine (continued)	
gas safety	85-88
laser	98-99
Fused silica	
properties	124
transmission	125, see Laser optics
Gain meduim	7, 68
Gain profile	70, 264, 270
Gas (excimer)	74
Gold reflection	140
Homogenizer	147, 234-235
Illumination	255
Index of refraction	
of optical materials	124
caluclation	130
Krypton fluoride	see Excimer laser
Lamps	11-16
Laser	
ablation	33-46
alexandrite	24-26
amplification	5-6
angioplasty	320-322
annealing	210-217
argon-ion	22
beam profiles	76-79
beam splitters	148
cleaning	107-109, 173-205
copper vapor	22
damage	126-127, 154-158
deposition	103
discharge chamber	68, 71-73
doping	217-219
excimer	67-93
gain medium	7, 68
gain profile	70, 264, 27

Laser (continued)
- gas — 48, 74, 79-80
- helium-cadmium — 23
- helium-neon — 23-24
- history — 3, 67-68
- imaging — 53, 104-107, 251
- inert gas — 21
- krypton-ion — 22
- line narrowing — 264-266
- lithography — 53-54, 261-270
- maintenance — 270-272
- micromachining — 287
- nitrogen — 23
- operation — 9-10
- optics — 8, 68, 124, 137-139
- planarization — 220-245
- preionizer — 72
- pulses — 76-79, 81
- quadrupled YAG — 18-21
- reliability — 80
- resonator — 8
- safety — 85-88
- solid state — 18, 24-26
- spectral analysis — 95-98
- spectrum — 2
- sub-systems — 71-75
- surface analysis — 104
- theory of operation — 4-9
- tunable — 101-102
- wavelengths — 48

Lithium fluoride — 124-125

Magnesium fluoride — 126, 144-145
Medical applications — 317-318; 320-333
Mercury lamp — 260
Microforming
- metal parts — 302-304
- process of — 300-306
- resist images — 304-306

Index 345

Micromachining
    catheters   332-333
    ceramic   297-299
    collagen   319
    copper   293-295
    glass fibers   308, 310
    micromoters (MEMS)   306-308
    optics   296
    plastic   293
    processes   288-290
    sapphire   299-300
    technology   287-288
Metal vapor lasers   22-23
Microbiology
    laser machining   327-332
    cutting of blood cells   327-328
Microelectromechanical systems (MEMS)
    process of   287-288, 306-308
    example of products   290-291, 301
Microelectronics   see Microlithography
Microlithography
    history   251-254
    illumination   255
    imaging principles   254-256
    lasers   261-266
    line narrowing   264-269
    optics   257-259
    photoresists   104-106
    spectral narrowing   264-269
    wafer stepper   252-253, 257
Microsurgery   327-332
Mid-ultraviolet   2
Mirror lens   292
Multi-layer dielectric coatings
    theory of   136
    structure   137
    reflectance curves   137-140
Multi-photon absorption   See Laser damage

Near ultraviolet   3

Nd:YAG 18-21

Optics
    ablation with 332
    attenuation 146-147
    beam delivery 124; 332
    coatings 123-151
    damage 126-129, 145, 151-154, 157-158
    efficiency 132-133
    excimer laser 85
    fiber beam delivery 332
    homogenizer 147
    lens materials 124, 259-260
    lithographic 258
    reflection 130
    transmission 125

Optical delivery systems
    ablation 292
    efficiency 132-133
    fibers 158-159, 321
    lamps 14-17
    laser 261, 266
    lithography 252-253, 258
    loss 132-133
    micromachining 292, 297, 329, 332
    planarization 231, 234
    scanning deposition 243
    tunable laser 102
    wavemeter 268
    xuv 100

Optical fiber 132-133
Optical path see Optical delivery
Organosilicons 107
Output power see Laser energy

Pattern generator 252
Photoablation
    characteristics 46-54
    ceramic 297-299

Index 347

Photoablation (continued)
- history — 44-46
- mechanisms — 33-34, 36-40
- metals — 42, 293-297
- parameters — 44-46
- plastics — 292-294
- of the human eye — 325-327
- plumes — 38-39
- theory — 33-36
- thresholds — 41-44

Photoacoustical signals — 45
Photoionization — 38
Photoluminescence — 38
Photolytic reactions — see Photoablation
Photon
- energy — 51
- theory — 4-8

Photoreactive cleaning — 109
Photorefractive keratectomy — 325
Photoresist
- ablation — 290
- for lithography — 106-107, 256

Planarization
- aluminum — 229-230, 235-244
- equipment — 231-233
- optics — 233-234
- process — 239-241
- semiconductor — 235-244
- theory — 220-230
- with polymers — 223-226

Plumes — 38-39, see Photoablation
PMMA — 42-43
Polarization — 156-157
Polymers
- ablation of — 39, 43, 290, 293
- planarization of — 221-226
- research — 104-107

Population inversion — 6-8
Preionization — 71-72

Printed circuits	191-195
Projection exposure	see Optical delivery
Proximity printing	
Pulses	
length	322-324
energy, horizontal/vertical	76-77
profile	78
variation	79
versus laser gas fill	83
Pyrex	125
Quadrupled YAG	18-21
Quartz	
properties of	126-127
transmission	125
Radar	3
Reflection	130-134
Reflective coatings	138-145; 149-151
Refractive index	124
Refractive lens	258
Resolution	254-256
Safety	85-88
Sapphire machining	299-300
Schwarzchild lens	292
Shockwave	45
Silicon dioxide	
ablation	37
coatings	143-144
Silver reflection	140
Solarization	see Laser damage
Solid state lasers	18-26
Spectral absorption	38, 40-44
Spectral analysis	95-98
Spectral emission (uv lamps)	12-13
Spectral properties of optical materials	40-41
Spectral narrowing	261, 264-268
Spectroscopy	see Spectral analysis

Index 349

Spectrum
    deep uv   1-2
    xuv   131
Spontaneous emission   4
Stimulated emission   5

Thin film heads   181-183
Thorium fluoride   143
Threshold of ablation   145, 295
Topography   221-226, see Planarization
Two-Photon absorption   324

Ultraviolet
    ablation   36-41
    absorption   38-41
    angioplasty   320-325
    bondbreaking   37
    characteristics   46
    cleaning   173, see Chapter 6
    coatings   123, see Chapter 5
    damage   145, 154-157
    delivery systems   12-17
    extreme uv   131-132
    imaging   53-54, see Lithography
    lamps   11-16
    machining   287
    microforming   300-306
    mirrors   150
    optics   234, 243 258, 292, 296, 310
    photon energy   51
    resolution   53-54
    safety   85-88
    sources   10
    spectrum   1-2
    surface reactions   49
    transmission   125
    vacuum uv   2
    wavelengths   47-48, 51-52

Via holes   237-239

Vacuum uv	2
Wafer cleaning	177-180
Wafer stepper	253, 257-259
Wavelength	
exposure	256
lasing specie	48
photon energy	52
reflection	147-151
Wavemeter	267-269
Wet cleaning	179-181
X-ray	
generation with excimer laser	99-100
spectrum	1
Xuv coatings	131-132